CHEMISTRY RESEARCH AND APPLICATIONS

INDOLE

SYNTHESIS, FUNCTIONS AND REACTIONS

CHEMISTRY RESEARCH AND APPLICATIONS

Additional books and e-books in this series can be found on Nova's website under the Series tab.

CHEMISTRY RESEARCH AND APPLICATIONS

INDOLE

SYNTHESIS, FUNCTIONS AND REACTIONS

BRIDGET PRATT
EDITOR

nova
science publishers
New York

Library of Congress Cataloging-in-Publication Data

ISBN: 978-1-53614-777-3

Published by Nova Science Publishers, Inc. † New York

CONTENTS

PREFACE

In the foremost chapter of Indole: Synthesis, Functions and Reactions, derivatives of hydrogenated pyrido[4,3-b]indoles as potential neuroprotectors have been synthesized. Different substituents were introduced into position 8 of the carboline fragment of the molecule: methyl-, methoxy-, fluorine- and chlorine-ones. Biological tests have shown that all the studied compounds can modulate glutamate-dependent uptake of calcium ions in rats' cerebral cortex synaptosomes.

The authors go on to discuss how mushrooms and their mycelia from in vitro cultures have become increasingly common study subjects due to their health-promoting potential. The most frequently demonstrated properties of these mushrooms include antioxidative, anticancer, anti-inflammatory, immunostimulatory and anti-depressive effects. This activity stems from numerous mushroom metabolites, bioelements and physiologically active substances, including indole compounds.

The final chapter provides an overview of the diverse synthetic approaches to generate pyrimido[4,5-b]indoles. Pyrimido-indoles are tricyclic ring systems produced when an indole group is fused with a pyrimidine ring. Of the two possible isomers of this ring system, the pyrimido[4,5-b]indoles are of particular interest relative to pyrimido[5,4-b]indoles due to their presence in a wide range of pharmacologically active molecules.

Chapter 1 - Derivatives of hydrogenated pyrido[4,3-b]indoles as potential neuroprotectors have been synthesized. Different substituents were introduced into position 8 of the carboline fragment of the molecule: methyl-, methoxy-, fluorine- and chlorine-ones. Biological tests have shown that all the studied compounds can modulate glutamate-dependent uptake of calcium ions in rats' cerebral cortex synaptosomes. The effect of the chemical nature of substituents introduced into the tricyclic fragment of the molecules on the compounds ability to positively modulate the activity of ionotropic glutamate AMPA receptors been studied. It has been found that the unsubstituted derivative, compounds with a methyl group and a fluorine atom have the highest bioactivity in this test. The shake-flask method has been used to measure the solubility of the compounds in the buffer solution (pH 7.4), hexane and 1-octanol within the temperature interval of 293.15 – 313.15 K. All the derivatives have been found to have low solubility (not exceeding $8 \cdot 10^{-4}$ mole fractions) in the mentioned solvents. The effect of thermophysical and protolytic properties of the compounds on the solubility has been studied and thermodynamic functions of compounds dissolution in the solvents used have been calculated. The partition coefficients of the synthesized derivatives in the system 1-octanol/buffer pH 7.4 in the temperature range of 293.15 - 313.15 K have been measured. It has been found that as the lipophilicity of the compounds studied increases, the maximum biological activity is achieved at a higher concentration of the administered substance. The thermodynamic transfer functions of the studied substances from an aqueous medium to the organic phase have been calculated.

Chapter 2 - Mushrooms and their mycelia from *in vitro* cultures are increasingly common study subjects due to their health-promoting potential. The most frequently demonstrated activity of these mushrooms include antioxidative, anticancer, anti-inflammatory, immunostimulatory and anti-depressive effect. This activity stems from numerous mushroom metabolites, bioelements and physiologically active substances, including indole compounds. Indole compounds present in mushrooms can be divided into tryptamine derivatives with hallucinogenic properties and bioactive indole derivatives, without hallucinogenic properties. In numerous species of edible mushrooms, including species living on wood, both the fruiting bodies as well as mycelia from *in vitro* cultures were found to contain non-hallucinogenic indole derivatives: 5-hydroxy-L-tryptophan, serotonin, L-tryptophan, tryptamine, melatonin, 5-methyl-D,L-tryptophan, indole-3-acetic acid, and indole. Moreover, the influence of substrate enriching for mycelial cultures with specific substances, including element salts on the profile of indole compounds and their total content has also been confirmed. These studies have enabled standardization of the conditions of *in vitro* culture conducting in laboratory, so as to obtain the highest possible contents of indole compounds in their biomass. Moreover, the influence of heat processing, analogous to cooking, has also been examined (as a method of preparing meals based on mushrooms) on the content of indole compounds. It has been demonstrated that heat processing results in approximately two-fold decrease of the content of indole compounds, but mushrooms remain their good source. It has also been demonstrated that fruiting bodies of selected edible mushroom species and their *in vitro* cultures not only synthesize indole derivatives and accumulate them, but they also release these substances to the solution of artificial gastric juice and artificial intestinal juice under *in vitro* conditions. Thus, they may constitute good source of indole compounds for human organism. The obtained highest degree of release of indole compounds from *in vitro* cultures in conditions stimulating natural digestive processes is particularly significant, as it enables development of commercial cultures accumulating these physiologically significant compounds, and in the future obtaining of enriched fruiting bodies, which may be used to produce natural diet supplement. Additionally, it should be emphasized that the presence of indole compound such as L-tryptophan and 5-hydroxy-L-tryptophan in fruiting bodies of edible mushrooms species additionally enhances their dietary values, which is of particular significance in vegetarian diet, where mushrooms rich in indole derivatives may constitute an alternative to red meat. In summary, indole compounds are successfully determined in mushrooms that demonstrate a multi-directional therapeutic and dietary activity.

Chapter 3 - Pyrimido-indoles are tricyclic ring systems produced when an indole group is fused with a pyrimidine ring. Of the two possible isomers of this ring system, the pyrimido[4,5-*b*]indoles are of particular interest relative to pyrimido[5,4-*b*]indoles, due to their presence in a wide range of pharmacologically active molecules. Compounds containing the pyrimido[4,5-*b*]indole chemotype have demonstrated affinity towards multiple anticancer biological targets including receptor tyrosine kinases, thymidylate synthase, BET bromodomains, and tubulin. In addition, derivatives bearing this scaffold exhibit antiviral, diuretic, anti-opportunistic infection, and hematopoietic stem-cell expansion activities. Multiple synthetic approaches to construct this notable tricycle have been reported including the Nenitzescu reaction, Fisher-type cyclization, palladium catalysis, and aza-Wittig type reaction among others. This book chapter provides an overview of the diverse synthetic approaches to generate pyrimido[4,5-*b*]indoles.

In: Indole: Synthesis, Functions and Reactions ISBN: 978-1-53614-777-3
Editor: Bridget Pratt

Chapter 1

NEW DERIVATIVES OF HYDROGENATED PYRIDO[4,3-B] INDOLES AS POTENTIAL NEUROPROTECTORS: SYNTHESIS, BIOLOGICAL TESTING AND PHARMACEUTICALLY RELEVANT PROPERTIES

Svetlana Blokhina, Angelica Sharapova and Marina Ol'khovich
Institute of Solution Chemistry,
Russian Academy of Sciences, Ivanovo, Russia

ABSTRACT

Derivatives of hydrogenated pyrido[4,3-b]indoles as potential neuroprotectors have been synthesized. Different substituents were introduced into position 8 of the carboline fragment of the molecule: methyl-, methoxy-, fluorine- and chlorine-ones. Biological tests have shown that all the studied compounds can modulate glutamate-dependent uptake of calcium ions in rats' cerebral cortex synaptosomes. The effect of the chemical nature of substituents introduced into the tricyclic fragment of the molecules on the compounds ability to positively modulate the activity of ionotropic glutamate AMPA receptors been studied. It has been found that the unsubstituted derivative, compounds with a methyl group and a fluorine atom have the highest bioactivity in this test. The shake-flask method has been used to measure the solubility of the compounds in the buffer solution (pH 7.4), hexane and 1-octanol [1]within the temperature interval of 293.15 – 313.15 K. All the derivatives have been found to have low solubility (not exceeding $8 \cdot 10^{-4}$ mole fractions) in the mentioned solvents. The effect of thermophysical and protolytic properties of the compounds on the solubility has been studied and thermodynamic functions of compounds dissolution in the solvents used have been calculated. The partition coefficients of the synthesized derivatives in the system 1-octanol/buffer pH 7.4 in the temperature range of 293.15 - 313.15 K have been measured. It has been found that as the lipophilicity of the compounds studied increases, the maximum biological activity is achieved at a higher concentration of the administered substance. The thermodynamic transfer functions of the studied substances from an aqueous medium to the organic phase have been calculated.

1. Introduction

A number of the most important drug structures have been identified by analyzing the results of experimental bioscreening of a large set of substances, their medical applications and structural data (Rekker and Mannhold, 1992). Among them are pyrido[4,3-b]indole derivatives with a wide range of biological activity: antihistamine, central depressant, anti-inflammatory, neuroleptic and other types (Ivachtchenko et al., 2010; Stefek et al., 2011; Dondas et. al., 2016). The structural analogues of this series, and the antihistamine drug Dimebon (Latrepirdine) in particular, were found to slow down the neurodegenerative process (Ikonomidou and Turski, 2002; Doody et al., 2008). Although Dimebon had not passed the final clinical tests, it stimulated new research aimed at studying molecular mechanisms of its activity and developing its efficient structural analogues (Eckert et al.

[1] Corresponding Author Email: avs@isc-ras.ru.

2012; Steele and Gandy, 2013; Hung et al., 2015). This work also aims to synthesize new Dimebon derivatives.

Modern studies of the Alzheimer's disease pathogenesis consider the pathological form of amyloid peptide to be the key factor in the development of neurodegenerative processe in human brain. Neurodegenerative processes caused by β-peptide include calcium homeostasis disorder, higher levels of oxidative stress, potentiation of toxicity of excitatory neurotransmitter amino acids, and cell death initiation (Mattson et al., 1992). An abnormal increase in calcium ion concentration in body cells is known to trigger a series of degenerative processes accompanying body aging. Calcium homeostasis disorder is the basis of the calcium theory of aging and dementia (Khachaturian, 1994). Scientists are now working on a whole range of approaches to prevention and correction of cell neurodegeneration including development of blocking agents of glutamate-induced calcium ion influx as potential neuroprotectors with a high biological activity (Parsons et al., 1998; Farlow, 2004). Another important area of research into new potential neuroprotectors is the study of their effect on AMPA receptors. The aim of activating this subtype of ionotropic glutamate receptors of the central nervous system is to improve cognitive functions in patients with neurodegenerative diseases.

Drug delivery plays an important role in creating biologically active compounds, along with specific interactions with ferments and receptors. A medicinal drug is expected to have a number of physical and chemical properties ensuring its distribution in the body. Among the most important properties is drug solubility in pharmaceutically relevant media that influences the drug ability to cross biological barriers, including the blood-brain barrier. Physicochemical properties of medicinal drugs, their structure and experimental data about solubility are now widely used to develop new methods of predicting solubility of biologically active compounds (Avdeef, 2007; Faller, 2007). But, despite the considerable progress in modeling methods, the accuracy of calculating the solubility of compounds with a large molecular weight and containing several functional groups remains rather low (Du-Cuny, 2008).

The main properties of drugs affecting their intestinal permeability are solubility in aqueous media and lipophilicity. It is quite natural because the substance can be absorbed through the intestine wall, if, on the one hand, it is well soluble in the gastrointestinal tract media and, on the other hand, it is lipophilic enough to permeate through the lipid membrane bilayer by means of passive diffusion. The degree of substance lipophilicity is calculated using the logarithm of the partition coefficient in the system 1-octanol/water (log*P*) (Sangster, 1997). This parameter is affected by compound ionization, acid-base properties, ability to form hydrogen bonds, charge distribution, etc.

Traditional methods of finding effective drug compounds consist in synthesizing a number of substances and their subsequent testing for different types of activity. A standard method characterizing the connection between the molecular structure of a substance, one of its properties and its biological activity is QSAR (Wang et. al., 2015). The log*P* value is the most frequently used descriptor for determining the quantitative relations "structure-activity" (Caron et al., 1999). There are now a lot of programs accurately determining this parameter based on the molecular structure (Mannhold et al., 2009). However, the accuracy of lipophilicity calculations for compounds with several functional groups remains insufficient.

Among such complex molecular structures are representatives of the γ-carboline series – derivatives of hydrogenated pyrido[4,3-b]indoles. Synthesis of new substances and revealing of the dependence of their biological activity on the substance physicochemical properties and molecular structure is the basis of the research direction named QSAR (Quantitative Structure-Activity Relationship) in modern medicinal chemistry (Danishuddin and Khan, 2016). To date, it is largely unknown how different functional groups introduced into a molecule change the substance influence on the body (Waterbeemd and Rose, 2008; Asirvatham et al., 2016). Solving this problem could help optimize the molecular design and develop efficient drugs with predictable biological activity. The aim of the work was to synthesize heterocyclic derivatives of tetrahydro-1H-pyrido[4,3-b]indole with substituents of different chemical nature in the carboline fragment, to study their neurocorrection activity, thermophysical properties, solubility and partition in pharmaceutically relevant media.

2. Experimental Section

2.1. General

All the chemicals were of reagent grade, purchased from commercial sources and used without further purification. Dimebon with 98% purity was purchased from Sigma-Aldrich (Saint Louis, MO, USA). The solvents used, namely 1-octanol and hexane from Merck, were of the highest purity. The ^{1}H and ^{13}C NMR spectra were recorded on a Bruker CXP-200 (Germany) (200.13 and 50.04 MHz, respectively). The chemical shift values are given in the δ scale relative to Me_4Si. The mass spectra were recorded on a Finnigan 4021 mass spectrometer operating at 70 eV. The elemental analyses were done on a Carlo-Erba CHN analyzer.

2.2. Biological Tests

2.2.1. Study of Glutamate- Induced $^{45}Ca^{2+}$ Uptake

The effect of the compounds on the glutamate-induced $^{45}Ca^{2+}$ uptake into synaptosomes of a rat brain cortex was determined by the method presented in the work (Hung, 2015). The interaction between the compounds and the glutamate-dependent calcium uptake system was studied on a P_2-fraction of mitochondrial crude synaptosomes (Gylys, 2000) obtained from the celebral cortex of newborn (9 – 10 days) Wistar rats using the standard Hajos method (Hajos, 1975). The brain was homogenized with 10 volumes of cooled 0.32 M sucrose at 900 rpm. The homogenate was centrifuged at 1500 g for 10 min, and the resulting supernatant was centrifuged at 10000 g for 20 min. The radioactive label was accumulated by a suspension of the synaptosomals P_2-fraction in incubation buffer A having the following composition: 132 mM NaCl, 5 mM KCl, 5 mM HEPES, 10 mM glucose, pH 7.4 (with the protein concentration of approximately 1.5-2 mg/ml). The biological activity of the synthesized derivatives was studied by adding 50 μl of buffer A to the incubation medium with the substance studied in the concentration range of 0.01−100 μM. The mixture was further incubated for 5 min with 200 μM of glutamate at 37°C and then the $^{45}Ca^{2+}$ uptake was stopped by filtration through GF/B-filters (Whatman, England) followed by three washes with cold buffer B (145 mM KCl, 10 mM Tris, 5 mM Trilon B, pH 7.4). The radioactivity probes were analyzed by a fluid scintillation β-analyzer (TriCarb, Perkin Elmer). All the tests were performed in three parallel trials in 2-3 independent experiments. The $^{45}Ca^{2+}$ uptake amount was estimated by calculating the difference between the radioactive label during stimulation with glutamate and without stimulation with an agonist. The result is presented as percentage in relation to the control measurement which was 100%. NMDA-antagonist Dizocilpine (MK-801) was used for positive Control. The MK-801 with 98% purity was purchased from Sigma-Aldrich (Saint Louis, MO, USA). The specific $^{45}Ca^{2+}$uptake was estimated using the following equation:

$$\text{Percentage } ^{45}Ca^{2+} \text{ of Control} = [(Ca_4\text{-}Ca_3)/(Ca_2\text{-}Ca_1)]\cdot 100\% \qquad (1)$$

where Ca_1- $^{45}Ca^{2+}$ is the uptake for the Control (without glutamate and the tested compound), Ca_2 is the glutamate-induced $^{45}Ca^{2+}$ uptake (glutamate only), Ca_3- $^{45}Ca^{2+}$ is the uptake in the presence of the tested compound (without glutamate), Ca_4- $^{45}Ca^{2+}$ is the uptake in the presence of glutamate and the tested compound.

2.2.2. Electrophysiological Patch-Clamp Method

The analysis of the testing compounds effects on AMPA receptors was carried out using the electrophysiological patch-clamp method in the whole cell configuration (Molleman 2003) of the Purkinje cell fresh isolated from the cerebellum of juvenile rats (12-15 days). The isolation was carried out using a modified method (Kaneda et al., 1988) in which the 400-600 μm cerebral sections were incubated for 60 min in a temperature controlled chamber in 10 mL of the buffer solution (150 mM NaCl,5 mM KCl, 2 mM $CaCl_2$ 2 mM $MgSO_4{\cdot}7H_2O$, 10mM HEPES, 15 mM glucose, pH 7,42). The incubation solution was replaced with the incubation buffer with pronase (2 mg/mL) and collagenase (1 mg/mL) for 70 min. After a 20 min wash with the original buffer, all the sections were placed into a Petri dish and were separated using a Pasteur pipette. All the solutions were constantly acrated with 100% O_2 at 34°C. The extracellular saline composition of was: 150 mM NaCl, 5 mM KCl, 2.6 mM $CaCl_2$, 2 mM $MgSO_4{\cdot}7H_2O$, 10mM HEPES, 15 mM glucose, pH 7,36. The transmembrane current was induced as a result of the activation of AMPA receptors by superfusion with the agonist solution (kainic acid). The currents were measured using borosilicate microelectrodes (with the resistance of 2.5-4.0 Ohms) filled with 100 mM KCl, 11 mM EGTA, 1mM $CaCl_2$, 1 mM $MgCl_2$, 10 mM HEPES, 5 mM ATR, pH 7.2. The currents were registered using a EPC-9 (HEKA, Germany) and recorded with the Pulse software package which was also purchased from HEKA.

2.3. Differential Scanning Calorimetry

The fusion temperatures and enthalpies of the compounds under investigation were determined using a Perkin-Elmer Pyris 1 DSC differential scanning calorimeter (Perkin-Elmer Analytical Instruments, Norwalk, Connecticut, USA) with the Pyris software for Windows NT. The DSC runs were performed in an atmosphere of flowing 20 cm^3/min dry helium gas of high purity 0.99996 (mass fraction) using standard aluminum sample pans at a heating rate of 2°C/min. The accuracy of the weight measurements was 0.005 mg. The DSC was calibrated using two-point calibration, measuring the onset temperatures of the indium and zinc standards. The melting onset was used for calibration because it is almost independent of the scan rate. The melting temperatures for indium and zinc were 156.6 and 419.5°C, respectively (determined by at least ten measurements). The enthalpy scale was calibrated using the indium fusion heat. The measured value of the fusion enthalpy corresponded to 28.69 J/g (the reference value was 28.66 J/g

(Archer et al., 2003). The standard uncertainty of the melting temperature was determined as twice the standard deviation of five independent measurements.

2.4. Solubility Experiment

The saturated equilibrium solubility of the studied compound was determined by the shake flask method at atmospheric pressure in the temperature range from 293.15 K to 313.15 K. The essence of the above mentioned method consists in determination of the compound concentration in the saturated solution. An excess amount of the compound was added to the known amount of various pure solvents in triplicate. The concentrated compound suspensions in each solvent were shaken continuously in an air thermostat containing a stirring device. The point of the solution thermodynamic equilibrium was determined based on the solubility kinetic dependences and averaged 72 hours. The solid phase sedimentation time after stirring was 4 hours. After the saturation was achieved, the solution aliquot was taken and centrifuged in a centrifuge Biofuge stratus (Germany) for 5 minutes at a fixed temperature. Then, the supernatant solutions were filtered through a 0.45 μm filter MILLEX®HA (Ireland). The saturated solution was diluted with the corresponding solvent to the required concentration. The concentration in molarity was assessed by measuring the absorbance at the wavelength of 329 nm using a UV-vis spectrophotometer (Cary-50, USA) at room temperature.

The calibration procedure was conducted at room temperature using the solutions with the known concentrations of each substance in each investigated solvent. The solutions were prepared by adding the appropriate substance mass and solvent (1-octanol, hexane) volume to the flask and mixing until the substance was totally dissolved. The absorbance of the solutions was measured and the calibration curves were constructed. The calibration curve appeared to be linear in the concentration range of 0.2–1.0 $\mu g \cdot ml^{-1}$ with a correlation coefficient of 0.9995. The experimental setup and its accuracy were validated by comparing the experimental solubility data of benzoic acid in water with those in (H. Stephen and T. Stephen, 1963). The deviation of the measured solubilities from the literature values did not exceed 1%.

The standard Gibbs energies of the dissolution processes ΔG^0_{sol} were calculated using the following equation:

$$\Delta G^0_{sol} = -RT \ln a_2 \tag{2}$$

where $a_2 = \gamma_2 x_2$ is the activity of the solute molecule; x is the drug molar fraction in the saturated solution; γ_2 is the activity coefficient of the solute molecule. The standard solution enthalpies ΔH^0_{sol} were calculated using the van't Hoff equation:

$$\partial(\ln a_2)/\partial T = \Delta H^0_{sol}/RT^2 \tag{3}$$

Since the solubilities of the compounds under investigation in buffer solutions were very low, it was assumed that $\gamma_2 = 1$. The temperature dependences of drug solubilities within the chosen temperature interval can be described by the linear function:

$$\ln x = \mathrm{A} - \mathrm{B}/T \tag{4}$$

This indicates that the change in the heat capacity of the solutions with the temperature is negligibly small.

The standard solution entropies ΔS^0_{sol} were obtained from the well-known equation:

$$\Delta G^0_{sol} = \Delta H^0_{sol} - T\Delta S^0_{sol} \tag{5}$$

2.5. Partition Experiment

The experiments for determination of the partition coefficients of the compounds in a 1-octanol/buffer system were carried out by the isothermal saturation method at five temperatures: 293.15, 298.15, 303.15, 308.15 and 313.15 K. The procedure was as follows: in an ampoule placed in a thermostat, we added a 1-octanol saturated buffer of a definite compound concentration to the identical volume of a buffer saturated 1-octanol solution. The resulting solution was equilibrated for two days with continuous shaking. The drug concentrations in both phases were determined by a Cary-50 spectrophotometer (USA) in the UV spectral region (λ = 190 – 400 nm) with an accuracy of 2 – 4%. The reported experimental values represent the average of at least three replicated experiments.

The ratio of the equilibrium molar concentration of the compound neutral molecules in the 1-octanol phase (C_o) to those in the water phase (C_b) was determined as the 1-octanol/buffer partition coefficient in the following form:

$$P = C_o/C_b \tag{6}$$

To calculate the thermodynamic functions of the transport process, the molar concentrations were recalculated into mole fractions x_o and x_b, respectively. The standard Gibbs energy of transfer $\Delta_{tr}G^o$ from buffer to organic systems was calculated as:

$$\Delta_{tr}G^o = -RT\ln P^* \tag{7}$$

where $P^* = x_o/x_b$

The partition temperature dependence (van't Hoff method) was employed to obtain the enthalpy of transfer $\Delta_{tr}H^o$:

$$\frac{d(\ln P^*)}{dT} = \frac{\Delta_{tr}H^o}{RT^2} \qquad (8)$$

The entropy of transfer $\Delta_{tr}S^o$ can be calculated from the following equation:

$$\Delta_{tr}S^o = (\Delta_{tr}H^o - \Delta_{tr}G^o)/T \qquad (9)$$

3. Results and Discussion

3.1. Synthesis and Characterization of Compounds

The synthetic goal of the work was to obtain new compounds of the carboline series – "direct analogues" of Dimebon possessing neurocorrecting activity. Dimebon modification consists in introducing a triazole cycle instead of the pyridyl one. The target compounds G4 were prepared using the synthetic route shown in Scheme I.

The pyrido[4,3-b]indoles (**G1**) were prepared by the Fischer reaction. The pyrido [4,3-b] indoles G1 were attached to the ethyl acrylate in the presence of Triton B in boiling benzene to give esters G2. The latter were converted by the hydrazine hydrate to the corresponding hydrazides G3. The reaction of the hydrazides of carboxylic acids and methyl isothiocyanate in boiling ethanol leads to 2,4-dihydro-1,2,4-triazole-3-thiones G4. The structures of these synthesized compounds were characterized by the ^{1}H NMR, ^{13}C NMR, MS methods and elemental analysis. In the ^{1}H NMR spectra in the 9-11 ppm range, there are no signals of NH-group protons of intermediate thiosemicarbazides, but a signal is observed in the range 13.37-13.54 ppm characteristic of 1,2,4-triazoles. The absence of the SH group proton signal in the NMR spectra indicated that the synthesized 1,2,4-triazoles were present in the thione form.

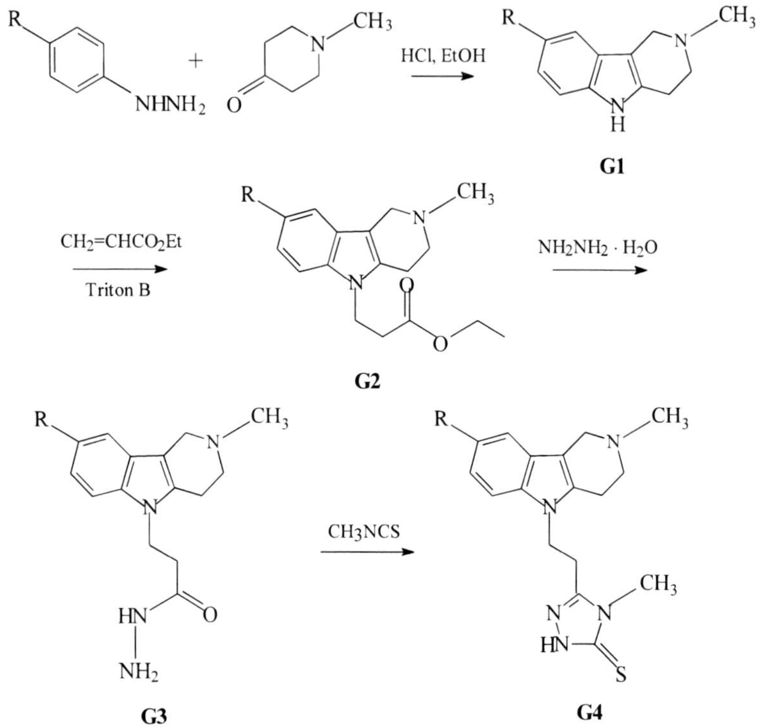

Scheme 1. Synthesis of 2,4-dihydro-1,2,4-triazole-3-thiones: R – H (I); $-CH_3$ (II); $-OCH_3$ (III); -F (IV); -Cl (V).

3.1.1. Synthesis of 3-(2-Methyl-2,3,4,5-Tetrahydropyrido[4,3-B] Indol-5-Yl) Propionic Acid Ethyl Ester (G2)

A solution of benzyltrimethylammonium hydroxide (40% wt. in methanol, 3.5 ml) was added to a solution of **G1** (17.114 g, 92 mmol) containing a mixture of benzene (170 ml) and ethyl acrylate (75 ml). The mixture was heated under reflux for 7 hours, cooled, washed with water, and extracted with 2M HCl. The acidic solution was extracted with benzene, basified with 10% NaOH, and extracted with benzene. The combined extracts were washed with water, dried and evaporated to an oil (21.18 g, 80%).

3.1.2. Synthesis of 3-(2-Methyl-2,3,4,5-Tetrahydropyrido[4,3-B]Indol-5-Yl)Propionic Acid Hydrazide (G3)

Hydrazine monohydrate (20 mL, 0.41 mol) was added to a solution of **G2** (20.356 g, 0.071 mol) in ethanol (50 mL), and the mixture was heated under reflux for 2h 30min. After 72-hour storage at -12 °C, the resulting crystals were collected, washed with 2-propanol, and dried to produce 12.7 g (65.6%) of **G3**, mp 123-124 °C. ^{1}H NMR (DMSO-d_6/$CDCl_3$, 1:1, δ, ppm, J/Hz): 9.07 (s, 1H, NH), 7.32 (m, 2H, H_{arom}), 7.01 (m, 2H, H_{arom}), 4.30 (t, 2H, J=6.7), 4.05 (br s, 2H, NH_2), 3.54 (s, 2H, $MeNCH_2$), 2,75 (m, 4H, $MeNCH_2CH_2$), and 2.47 (m, 5H, CH_2, Me).

The other acid hydrazides **G3** were prepared by the above-mentioned method for compound **I**.

The 3-(2,8-Dimethyl-1,2,3,4-tetrahydropyrido[4,3-b]indol-5-yl) propionic acid hydrazide was prepared in 73% yield, mp 163.5-165.5°C. ^{1}H NMR ($CDCl_3$, δ, ppm, J/Hz): 7.14 (m, 2H, H_{arom}), 7.02 (br, s, NH), 6.95 (dd,1H, H_{arom}, J=8.6, J=1.2), 4.31 (t, 2H, CH_2, J=6.6), 3.59 (s, 2H, $MeNCH_2$), 3.48 (br s, 2H, NH_2), 2.78 (s, 4H,$MeNCH_2CH_2$), 2.51 (s, 3H, Me), 2.42 (m, 5H, CH_2, Me).

The 3-(8-Methoxy-2-methyl-1,2,3,4-tetrahydropyrido[4,3-b]indol-5-yl) propionic acid hydrazide was prepared in 44.6% yield, mp 126.5-127.5°C. ^{1}H NMR (DMSO-d_6/$CDCl_3$, 1:1, δ, ppm, J/Hz): 9.05 (s, 1H, NH), 7.23 (d, H_{arom}, J=8.6), 6.79 (d, 1H, H_{arom}, J=2.2), 6.69 (dd, 1H, H_{arom}, J=8.6, J=2.2), 4.25 (t, 2H, CH_2, J=7.0), 3.76 (s, 3H, MeO), 3.51 (s, 2H, $MeNCH_2$), 2.79 (m, 4H, $MeNCH_2CH_2$), 2.42 (m, 5H, CH_2, Me).

The 3-(8-Fluoro-2-methyl-1,2,3,4-tetrahydropyrido[4,3-b]indol-5-yl) propionic acid hydrazide was prepared in 72% yield, mp 158-159°C. ^{1}H NMR (DMSO-d_6/$CDCl_3$, 1:1, δ, ppm, J/Hz): 9.06 (s, 1H, NH), 7.35 (dd, 1H, H_{arom}, J=8.6, J=4.5), 7.00 (dd, 1H, H_{arom}, J=9.6, J= 2.5), 6.84 (td, 1H, H_{arom}, J=9.2, J=2.5), 4.29 (t, 2H, J=6.7), 4.06 (br s, 2H, NH_2), 3.49 (s, 2H, $MeNCH_2$), 2.79 (m, 4H, $MeNCH_2CH_2$), 2.46 (m, 5H, CH_2, Me).

The 3-(8-Chloro-2-methyl-1,2,3,4-tetrahydropyrido[4,3-b]indol-5-yl) propionic acid hydrazide was prepared in 74% yield, mp 172-173°C. ^{1}H NMR (DMSO-d_6/$CDCl_3$, 1:1, δ, ppm, J/Hz): 9.06 (s,1H, NH), 7.35 (d, 1H, H_{arom}, J=8.6), 7.29 (d, 1H, H_{arom}, J=1.9), 7.02 (dd, 1H, H_{arom}, J=8.6, J=1.9),

4.29 (t, 2H, CH_2, J=6.7), 4.01 (br S, 2H, NH_2), 3.50 (s, 2H, $MeCH_2$), 2.80 (m, 4H, $MeNCH_2CH_2$), 2.46 (m, 5H, CH_2, Me).

3.1.3. Synthesis of 4-Methyl-5-[2-(2-Methyl-1,2,3,4-Tetrahydropyrido [4,3-B]Indol-5-Yl)Ethyl]-2,4-Dihydro[1,2,4]Triazole-3-Thione (G4 - Compound I)

A mixture of 2.187 g (8 mmol) 3-(2-methyl-1,2,3,4-tetrahydro-pyrido[4,3-b]indol-5-yl)-propionic acid hydrazide, 0.587 g (8 mmol) methyl isothiocyanate and 55 ml of ethanol was heated under reflux for 6 h 30 min and then left for 48 h at ambient temperature. The resulting crystals were collected, washed with ethanol, and dried. Yield 2.268 g (86%), mp 216-219°C (EtOH). Found: C 62.58; H 6.54; N 21.27%. Calcd for $C_{17}H_{21}N_5S$: C 62.36; H 6.46; N 21.39; S 9.79%. ^1H NMR (DMSO-d_6/$CDCl_3$, 1:1, δ, ppm, J/Hz): 13.51 (br s, 1H in exchange, NH), 7.31 (m, 2H, H_{arom}), 7.04 (m, 2H, H_{arom}), 4.43 (t, 2H, CH_2, J=6.9), 3.56 (s, 2H, $MeNCH_2$), 3.15 (s, 3H, Me), 3.06 (t, 2H, CH_2, J=6.9), 2.75 (s, 4H, $MeNCH_2CH_2$), 2.48 (s, 3H, Me). ^{13}C NMR (DMSO-d_6, δ, ppm): 166.55, 150.21, 135.52, 133.26, 125.14, 120.40, 118.67, 117.22, 108.94, 107.54, 51.88, 51.15, 45.32, 29.32, 25.68, 21.82. MS (m/z): 327 (M^+).

The other 2,4-dihydro-1,2,4-triazole-3-thiones were prepared by the above-mentioned method for compound **I**.

5-[2-(2,8-Dimethyl-1,2,3,4-tetrahydropyrido[4,3-b]indol-5-yl)ethyl]-4-methyl-2,4-dihydro [1,2,4] triazole-3-thione (compound **II**). Yield 80%, mp 219-222 °C (EtOH). Found: C 63.52; H 6.73; N20.64%. Calcd for $C_{18}H_{23}N_5S$: C 63.31; H 6.79; N 20.51; S 9.39%. ^1H NMR (DMSO-d_6/$CDCl_3$, 1:1, δ, ppm, J/Hz): 13.37(br s, 1H in exchange, NH), 7.14 (m, 2H, H_{arom}), 6.90 (dd, 1H, H_{arom}, J=8.3, J=1.3), 4.39 (t, 2H, CH_2, J=6.9), 3.54 (s, 2H, $MeNCH_2$), 3.13 (s, 3H, Me), 3.03 (t, 2H, CH_2, J=6.9), 2.73 (m, 4H, $MeNCH_2CH_2$), 2.48 (s, 3H, Me), 2.39 (s, 3H, Me). ^{13}C NMR (DMSO-d_6, δ, ppm): 166.53, 150.25, 133.97, 133.26, 127.12, 125.36, 121.85, 117.03, 108.65, 107.04, 51.89, 51.18, 45.32, 29.32, 25.70, 21.83, 21.05. MS (m/z): 341 (M^+).

5-[2-(8-Methoxy-2-methyl-1,2,3,4-tetrahydropyrido[4,3-b]indol-5-yl)ethyl]-4-methyl-2,4-dihydro[1,2,4]triazole-3-thione (compound **III**).

Yield 93%, mp 219-222°C (EtOH). Found: C 60.71; H 6.54; N 19.47%. Calcd for $C_{18}H_{23}N_5OS$: C 60.48; H 6.49; N 19.59; O 4.48; S 8.97%. ^{1}H NMR (DMSO-d_6/$CDCl_3$, 1:1, δ, ppm, J/Hz): 13.51 (br s, 1H in exchange, NH), 7.18 (d, 1H, H_{arom}, J=8.8), 6.82 (d, 1H, H_{arom}, J=1.9), 6.70 (dd, 1H, H_{arom}, J=8.8, J=1,9), 4.39 (t, 2H, CH_2, J=6.7), 3.78 (s, 3H, MeO), 3.53 (s, 2H, $MeNCH_2$), 3.13 (s, 3H, Me), 3.03 (t, 2H, CH_2, J=6.8), 2.72 (m, 4H, $MeNCH_2CH_2$), 2.47 (s, 3H, Me). ^{13}C NMR (DMSO-d_6, δ, ppm): 166.51, 153.20, 150.27, 133.83, 130.66, 125.48, 109.78, 109.58, 107.28, 99.76, 55.21, 51.89, 51.26, 45.32, 29.30 25.75, 21.90. MS (m/z): 357 (M^+).

5-[2-(8-Fluoro-2-methyl-1,2,3,4-tetrahydropyrido[4,3-b]indol-5-yl)ethyl]-4-methyl-2,4-dihydro[1,2,4]triazole-3-thione (compound **IV**). Yield 83%, mp 215-218°C (EtOH). Found: C 59.39; H 5.75; N 20.38%. Calcd for $C_{17}H_{20}FN_5S$: C 59.11; H 5.84; F 5.50; N 20.27; S 9.28%. ^{1}H NMR (DMSO-d_6/$CDCl_3$, 1:1, δ, ppm, J/Hz): 13.49 (br s, 1H in exchange, NH), 7.26 (dd, 1H, H_{arom}, J=8.9, J=4.4), 7.03 (dd, 1H, H_{arom}, J=9.6, J=2.4), 6.84 (td, 1H, H_{arom}, J=9.2, J=2.4), 4.43 (t, 2H, CH_2, J=7.0), 3.52 (s, 2H, $MeNCH_2$), 3.16 (s, 3H, Me), 3.05 (t, 2H, CH_2, J=7.0), 2.75 (s, 4H, $MeNCH_2CH_2$), 2.48 (s, 3H, Me). ^{13}C NMR (DMSO-d_6, δ, ppm, *J*/Hz): 166.57, 156.82 (d, $^1J_{CF}$=231.4), 150.16, 135.47, 132.22, 125.32 (d, $^3J_{CF}$=9.9), 109.91 (d, $^3J_{CF}$=9.6), 108.06 (d, $^2J_{CF}$=22.1), 107.75, 102.33 (d, $^2J_{CF}$=23.2), 51.76, 50.99, 45.27, 29.34, 25.67, 21.95. MS (m/z): 345 (M^+).

5-[2-(8-Chloro-2-methyl-1,2,3,4-tetrahydropyrido[4,3-b]indol-5-yl)ethyl]-4-methyl-2,4-dihydro[1,2,4]triazole-3-thione (compound **V**). Yield 88%, mp 216-219 °C (EtOH). Found: C 56.17; H 5.64; N 19.47%. Calcd for $C_{17}H_{20}ClN_5S$: C 56.42; H 5.57; Cl 9.80; N 19.35; S 8.86%. ^{1}H NMR (DMSO-d_6/$CDCl_3$, 1:1, δ, ppm, J/Hz): 13.54 (br s, 1H in exchange, NH), 7.32 (d, 1H, H_{arom}, J=1.9), 7.27 (d, 1H, H_{arom}, J=8.7), 7.02 (dd, 1H, H_{arom}, J=8.7, J=1.9), 4.43 (t, 2H, CH_2, J=6.8), 3.53 (s, 2H, $MeNCH_2$), 3.18 (s, 3H, Me), 3.05 (t, 2H, CH_2, J=6.8), 2.75 (s, 4H, $MeNCH_2CH_2$), 2.48 (s, 3H, Me). ^{13}C NMR (DMSO-d_6, δ, ppm): 166.56, 150.09, 135.32, 134.06, 126.19, 123.38, 120.10, 116.60, 110.54, 107.50, 51.70, 50.83, 45.24, 29.37, 25.62, 21.86. MS (m/z): 361 (M^+).

3.2. Biological Activity

3.2.1. Effect of Compounds on Glutamate-Induced $^{45}Ca^{2+}$ Uptake

In order to develop medicinal drugs with neurocorrecting activity, we synthesized derivatives of the hydrogenated pyrido[4,5-b]indoles (**I – V**). Their structural formulae are given in Scheme 1. Since the target of Dimebon and, consequently, of the obtained compounds is the glutamatergic system of the central nervous system, the *in vitro* studies of their physiological activity were conducted on the P_2-fraction of synaptosomes of rat celebral cortex containing NMDA, kainate and metabotropic receptors of type I. The synthesized compounds have been found to be able to modulate the glutamate-dependent uptake of calcium ions in rat cerebral cortex synaptosomes. The results of studying the biological activity of the synthesized compounds are given in Table 1 and Figure 1. As these data show, the concentration dependences of the inhibitory factor have a dome shape, just like those of Dimebon.

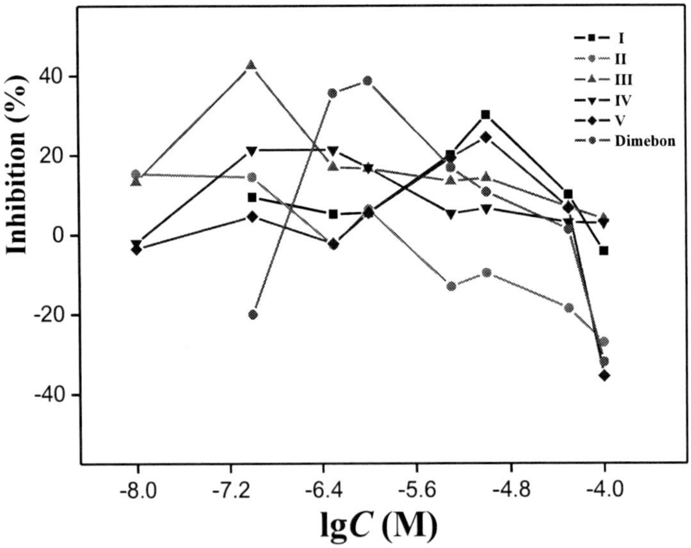

Figure 1. Glutamate-induced $^{45}Ca^{2+}$ uptake into rat cortex synaptosomes at different concentrations (lg*C*) of the studied compounds introduced into the incubation mixture of the suspension of synaptosomes.

Table 1. Effect of compounds studied on glutamate-induced $^{45}Ca^{2+}$ uptake into synaptosomes of rat cortex

Concen-tration, μM	Percentage $^{45}Ca^{2+}$ of Control (Control – 100%)						
	MK-801	I	II	III	IV	V	Dimebon
0.01	99.7	-	84.7 ± 1.2	86.7± 4.0	102.0 ± 9.8	103.5 ± 7.7	–
0.1	98.9	90.5±5.1	85.5 ± 3.9	57.5± 8.3	78.7 ± 10.3	95.3 ± 1.0	120.5 ± 3.3
0.5	98.4	94.8±3.0	102.6±3.6	83.0 ±7.9	78.7 ± 8.0	102.2 ± 6.3	64.4± 12.4
1	97.5	94.5± 1.8	93.6 ± 3.6	83.4± 2.7	83.2 ± 10.9	94.5 ± 4.7	61.3 ± 8.1
5	76.1	80.0 ± 3.0	113.1±7.6	86.6 ± 4.5	94.7 ± 8.5	80.7 ± 10.5	83.2 ± 4.1
10	65.2	69.9 ± 4.4	109.7±7.5	85.8±15.2	93.4 ± 6.9	75.6 ± 10.0	89.3 ± 1.2
50	35.0	90.0 ± 6.1	118.6±2.7	93.1±12.8	96.9 ± 6.0	93.3 ± 11.0	98.7 ± 8.8
100	26.9	104.2±12.8	127.1±17.1	96.1± 3.2	87.1 ± 11.5	135.5 ± 9.4	132.0 ± 2.9

Compound **I** did not affect glutamate-induced calcium ion uptake into rat cortex synaptosomes at the concentration of 0.1 – 1.0 μM. At the concentrations of 5 – 50 μM, this compound inhibited calcium ion uptake. The maximum inhibition was 30% at the substance concentration of 30 μM. Compound **II** inhibited glutamate-induced uptake of calcium ions into the synaptosomes (inhibition ~ 15%) at the concentration of 0.01 – 0.1 μM. At the concentrations of 5 – 100 μM, the studied compounds caused an increased uptake of calcium ions into the synaptosomes (109 – 127% of control). The inhibiting ability of compound **III** became apparent at the concentration of 0.01 – 10 μM, with the maximal calcium ion uptake observed at the concentration of 0.1 μM (42.5%). Compound **IV** acted as an inhibitor in this test at the concentration of 0.1 – 1 μM and had no effect at the concentration of 5 – 100 μM. The inhibition of glutamate-induced calcium ion uptake with compound **V** was most efficient (~ 20%) at the concentration of 5 – 10 μM. The increase in the calcium ion uptake into the rat cortex synaptosomes was observed at the concentration of 100 μM (135.5%). As the obtained results show, all the studied compounds have inhibiting activity in this test. And their biological activity depends on the chemical nature of the substituent introduced into position 8 of the tetra-hydro-1H-pyrido[4,3-b]indole fragment of the molecule. The highest inhibiting activity within a wide concentration range (0.01 – 10 μM) comparable to the Dimebon action (Table 1) was found in compound **III** with a methoxy-substitutent, which allows us to recommend it for further studies. Compounds **IV** and **V** with fluorine and chlorine atoms as substituents are weak inhibitors of calcium ion uptake, and methyl-derivative **II** even potentiates the uptake at high concentrations. Besides, by comparing the inhibiting ability of compound **II** and Dimebon (Figure 1), we can conclude that the biological activity of the studied compounds depends both on the substituent nature in the tetrahydro-carboline group and on the structure of the fragment bound through an ethyl linker.

3.2.2. Effect of Compounds on AMPA Receptors

Positive modulation of the ionotropic glutamate AMPA receptor activity in the central nervous system increases synaptic transmission, which results in memory formation and improvement and triggers the expression mechanism of genes responsible for synthesizing neurotrophins – growth factors regulating neuron survival and functioning (Shaffer et al., 2013). The effect of compounds on the functional activity of the AMPA-dependent ionic channels is an indicator of their potential cognitive stimulation properties (Lynch, 2006). The experiments with the synthesized Dimebon derivatives (**I – V**) were made on single isolated Purkinje neurons excreted from rat brain cerebellum. The results are shown in Table 2.

Table 2. Amplitudes of kainic acid-induced currents (A) in Purkinje cells at different concentrations (C) of the studied compound

C, mol/l	A, %				
	I	II	III	IV	V
10^{-10}	100 ± 9	100 ± 8	100 ± 7	100 ± 9	100 ± 8
10^{-9}	148 ± 11	100 ± 11	92 ± 11	126 ± 10	100 ± 5
10^{-8}	120 ± 9	100 ± 10	130 ± 8	135 ± 11	100 ± 7
10^{-7}	111 ± 8	117 ± 7	110 ± 9	150 ± 10	105 ± 7
10^{-6}	105 ± 7	148 ± 8	100 ± 8	100 ± 8	115 ± 7
10^{-5}	100 ± 6	90 ± 8	70 ± 11	90 ± 13	110 ± 8
$3 \cdot 10^{-5}$	81 ± 8	-	-	-	-

Note: the amplitudes of the control currents are taken as 100%.

Application of kainic acid facilitated generation of the inward transmembrane currents caused by AMPA-receptor activation. Adding compounds **I** – **IV** to the solution bathing the neuron led to concentration-dependent current amplitude changes as compared to the reference substance, while adding compound **V** did not have a great effect on the amplitude (Figure 2).

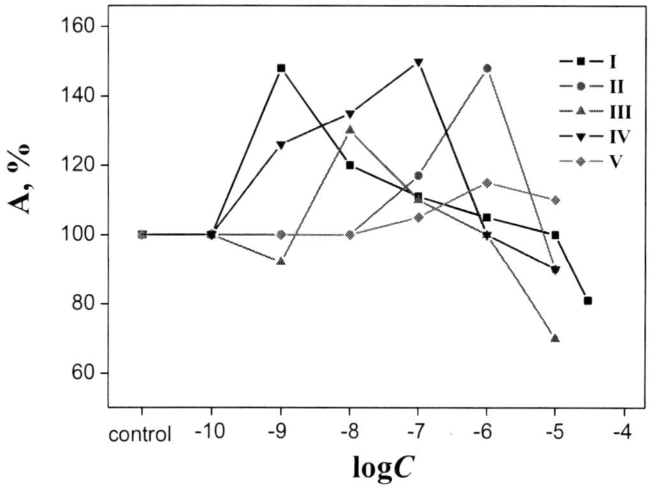

Figure 2. Amplitudes of kainic acid-induced currents (A, %) in Purkinje cells at different concentrations (*C*, mol/l) of the studied compounds.

The highest activity in this trial (~150% compared to the reference compound) was observed in unsubstituted compound I, in compound II with a methyl group and compound **IV** with a fluorine atom introduced into position 8- of the aromatic ring of the tricyclic molecule fragment. The bioactivity of compound **III** with a methoxy- group was a little lower. The maximum positive AMPA receptor modulation was achieved in the following concentrations of the introduced compounds: 10^{-9}, 10^{-6}, 10^{-8}, 10^{-7} mol/l for derivatives **I-IV**, respectively. The substance concentration increased up to 10^{-5} mol/l and higher if the potentiation decreased so much that there was no positive effect at all.

3.3. Thermophysical Properties

The thermophysical properties of the synthesized compounds were studied by the DSC method. The results of these studies are given in Table 3. The DSC curves shown in Figure 3 allow us to conclude that the compounds do not have polymorphic modifications and are stable when heated to melting temperatures.

The obtained melting temperature values are within the range of 234.1 – 276.0°C, while the melting enthalpies vary in the range of 39.7 – 53.6 kJ/mol. Derivative **II** with a methyl substituent has the highest melting temperature and enthalpy. Replacing the fluorine atom with a chlorine one in the molecular structure of compound **IV** and the transition to the structure of compound **V** result in a melting temperature increase but do not change the melting enthalpy.

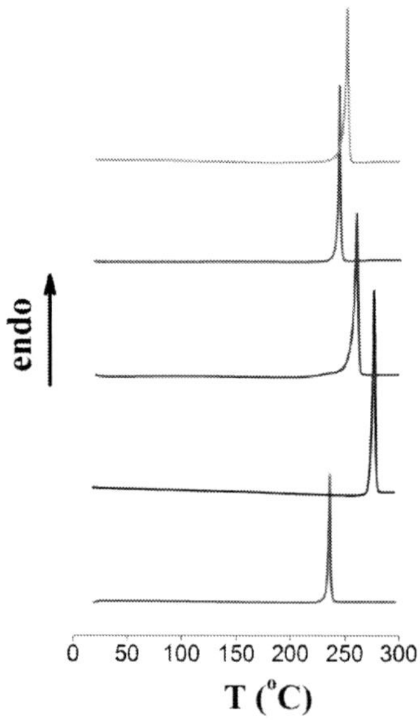

Figure 3. DSC-curves of the studied compounds.

Table 3. Physico-chemical characteristics of the studied compounds

Compound	[a]M (g/mol)	[b]T_m (°C)	[c]H_m (kJ/mol)	[d]pK_a
I	327.45	234.1 ± 0.2	44.3 ± 0.5	9.65 ± 0.2
II	341.46	276.0 ± 0.2	53.6 ± 0.5	9.63 ± 0.2
III	357.47	254.5 ± 0.2	39.7 ± 0.5	9.62 ± 0.2
IV	345.44	272.7 ± 0.2	41.1 ± 0.5	9.62 ± 0.2
V	361.89	245.9 ± 0.2	40.9 ± 0.5	9.61 ± 0.2

[a]M – molecular mass of compound,
[b]T_m – melting temperature of compound.
[c]H_m – melting enthalpy of compound.
[d]calculated using Advanced Chemistry Development (ACD/Labs) Software V11.02.

3.4. Solubility in Pharmaceutically Relevant Solvents

The classical saturation shake-flask method has been used to measure the solubility of the synthesized substances as it produces the most accurate results even for poorly soluble compounds (Baka et al., 2008). A buffer pH 7.4 simulating the blood plasma and 1octanol modeling the membrane lipid layer were used as solvents (Kerns and Di, 2008). In order to find out the capacity of the studied organic substances for specific interactions, we selected hexane as it interacts with the dissolved substance only through the van-der-Waals forces.

The solubility of the studied derivatives of hydrogenated pyrido[4,3-b]indoles (**I** – **V**) was measured in the buffer solution (pH 7.4), 1-octanol and hexane in the temperature range of 293.15 – 313.15 K (Table 4). Figure 4 shows the temperature dependences of compound **I** solubility as an example. Solubility of organic substances depends on whether the molecules are in the charged or neutral state in the water solution (Avdeef, 2007). The molecules of the studied compounds belong to the γ-carboline class, have a tetracyclic structure and contain an N-acceptor fragment (the main atom in the heterocycle is N) and an N-donor fragment (an acidic secondary amino group).

Table 4. Experimental mole fraction solubilities (x) of the tudied compounds in buffer pH 7.4, hexane and 1-octanol at different temperatures

T (K)	**I**			**II**			**III**		
	buffer pH 7.4	hexane	1-octanol	buffer pH 7.4	hexane	1-octanol	buffer pH 7.4	hexane	1-octanol
	$x \cdot 10^6$	$x \cdot 10^6$	$x \cdot 10^4$	$x \cdot 10^6$	$x \cdot 10^5$	$x \cdot 10^4$	$x \cdot 10^6$	$x \cdot 10^5$	$x \cdot 10^4$
293.15	1.69	7.96	2.16	1.64	1.42	0.89	5.57	3.30	0.71
298.15	1.86	9.41	2.74	1.96	1.70	1.13	6.55	3.71	0.93
303.15	2.04	11.41	3.48	2.27	2.02	1.39	7.53	4.18	1.22
308.15	2.23	13.26	4.28	2.70	2.35	1.68	8.77	4.68	1.61
313.15	2.42	15.50	5.40	3.21	2.72	2.11	10.09	5.31	2.05
A^a	-7.76 ± 0.04	-1.25 ± 0.18	5.83 ± 0.11	-2.90 ± 0.22	-1.39 ± 0.17	3.95 ± 0.23	-2.82 ± 0.09	-2.93 ± 0.14	7.10 ± 0.15
B^a	1649 ± 12	3073 ± 39	4179 ± 45	3054 ± 67	2989 ± 49	3889 ± 69	2719 ± 29	2165 ± 43	4876 ± 47
R^b	0.9999	0.9997	0.9998	0.9993	0.9996	0.9995	0.9998	0.9994	0.9999
σ^c	$0.2 \cdot 10^{-2}$	$0.9 \cdot 10^{-2}$	$0.8 \cdot 10^{-2}$	$1.1 \cdot 10^{-2}$	$0.8 \cdot 10^{-2}$	$1.2 \cdot 10^{-2}$	$0.4 \cdot 10^{-2}$	$0.7 \cdot 10^{-2}$	$1.2 \cdot 10^{-2}$
T (K)	**IV**			**V**					
	buffer pH 7.4	hexane	1-octanol	buffer pH 7.4	hexane	1-octanol			
	$x \cdot 10^6$	$x \cdot 10^4$	$x \cdot 10^4$	$x \cdot 10^6$	$x \cdot 10^4$	$x \cdot 10^4$			
293.15	8.34	3.85	2.51	2.66	5.42	1.15			
298.15	9.60	4.47	3.01	3.25	6.39	1.47			
303.15	11.11	5.11	3.68	4.01	7.47	1.91			
308.15	12.66	5.96	4.39	4.95	8.87	2.42			
313.15	14.60	6.81	5.30	6.07	10.22	3.01			
A^a	-2.96 ± 0.09	1.09 ± 0.1	3,65± 0.15	0.12 ± 0.16	2.45 ± 0.12	6.03± 0.15			
B^a	2558 ± 27	2624 ± 30	3501 ± 46	3799± 50	2922 ± 38	4423 ± 47			
R^b	0.9998	0.9998	0.9997	0.9998	0.9997	0.9998			
σ^c	$0.4 \cdot 10^{-2}$	$0.5 \cdot 10^{-2}$	$1.0 \cdot 10^{-2}$	$0.84 \cdot 10^{-2}$	$0.7 \cdot 0^{-2}$	$0.8 \cdot 10^{-2}$			

[a] parameters of the correlation equation: $\ln x = A - B/T$; [b] R – pair correlation coefficient; [c] σ – standard deviation.

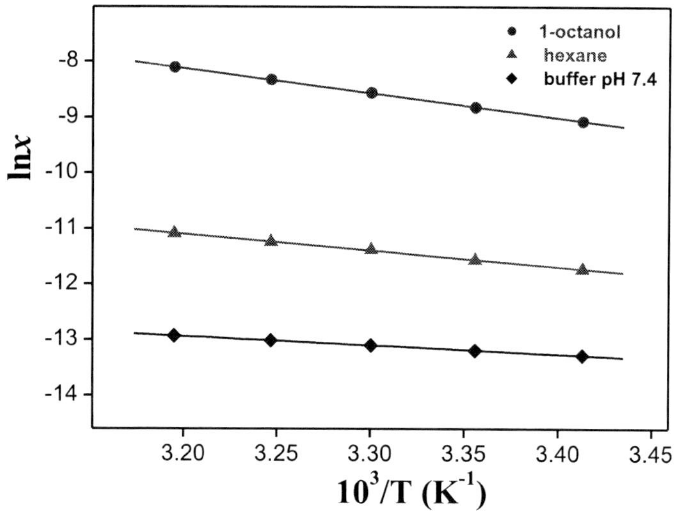

Figure 4. Temperature dependences of solubility of compound **I** in the selected solvents.

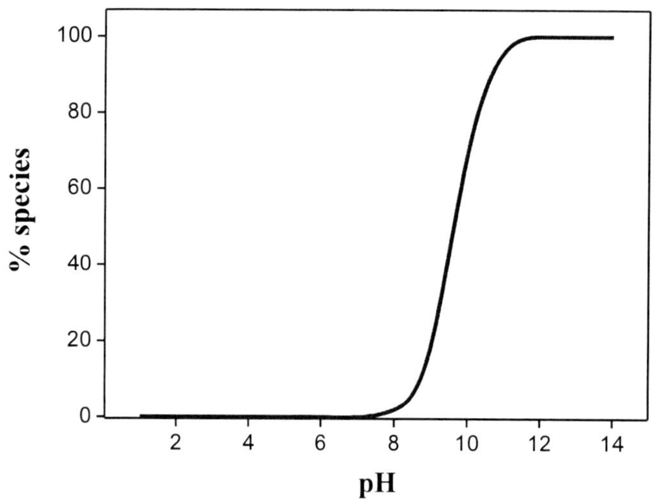

Figure 5. Distribution of the species as a function of the buffer solution pH for compounds I-V.

The dissociation constant values (Table 3) indicate that all the studied substances have almost the same protolytic properties and are strong acids, which is explained by the presence of an NH-group in the triazol-thion heterocycle. The content of ionized and nonionized molecule forms depending on the pH of the water solution was determined based on pK_a values using the Henderson-Hasselbach equation (Po et al., 2001). As Figure 5 shows, the molecules of the studied compounds are in the neutral state at pH < 7.0, while at pH > 7.0 they act as acids and become deprotonated. In the buffer solution (pH 7.4) all the molecules are in the neutral form.

All the studied compounds have very low solubility in the buffer (pH 7.4), with the solubility values ranging between $(1.86 - 46.2)\cdot10^{-6}$ mole fractions at the temperature of 298.15 K. As it has been determined, the solubility values of the substances can be arranged as follows: **I** < **II** < **V** < **III** < **IV** (Figure 6). The substitution of the hydrogen atom in the benzene ring of compound **I** for a methyl group and transformation of the latter into compound II slightly increase the solubility in the aqueous medium. In this case, the influence of the methyl group on the solubility of the mentioned heterocyclic compound results from this substituent's ability to disrupt the formation of hydrogen-bonded water associates and is not a direct effect of the electron density shift (Albert, 2007). Compounds **III** – **V** with halogen atoms and a methoxy-group as substituents have a comparatively higher solubility in the buffer.

The polar groups in these compounds probably produce an inductive effect shifting the electrons of the benzene ring to the atoms of fluorine, chlorine and oxygen, which results in stronger dipole-dipole interactions of the dissolved substances with the water molecules and higher solubility. The polar groups in these compounds probably produce an inductive effect shifting the electrons of the benzene ring to the atoms of fluorine, chlorine and oxygen, which makes the dipole-dipole interactions of the dissolved substances with the water molecules stronger and the solubility higher.

And the best solubility is observed in compound **IV** with a fluorine atom that is the most electronegative element.

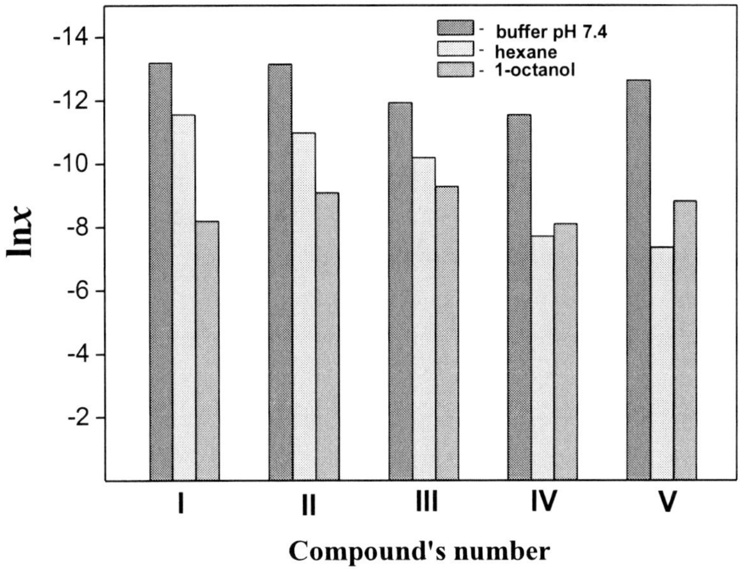

Figure 6. Solubility histogram of the studied compounds.

The solubility of the studied compounds in hexane is almost an order of magnitude higher than in the buffer solution but does not exceed the value of 4.87 10^{-4} mole fractions at the standard temperature. The only exception is unsubstituted derivative **I**, the solubility of which increases by 5 times only. Since hexane can interact with substances only through van-der-Waals forces, introduction of the substituents is characterized by higher solubility of the compounds due to induced dipoles formed in the molecule aromatic ring and the resulting attraction between the dissolved substance and the solvent. The electron donating methyl group of compound **II** increases the electron density of the aromatic ring and, consequently, the strength of interaction with hexane, which improves the solubility. The inductive effect of the oxygen atoms and halogens results in an electron shift toward the electron-accepting substituents. And it is quite natural that the substitution of the C-H bond for C-OCH_3, C-Cl and C-F bonds increases the solubility of the compounds as the atoms become more electronegative. Besides, it has been found that the solubility of derivatives **II** – **V** in hexane increases in the same order as their melting temperatures decrease (Table 3).

The data in Table 3 show that the solubility of the substances in 1-octanol is also very low, with the values between (0.71 – 5.40)·10^{-4} mole fractions in the studied temperature range. In the order of increasing solubility in this solvent, the compounds are arranged as follows: **III** < **II** < **V** < **I** < **IV**. And the solubility values are higher in 1-octanol than in the aqueous solution for all the compounds. The introduction of methyl- and methoxy- substitutents into the benzene ring insignificantly increases the solubility of compounds **II** and **III** in alcohol compared to hexane. The opposite effect is observed when the oxygen atoms are replaced with halogen ones in the benzene ring: the solubility values of compounds **IV** and **V** in 1-octanol are lower than in hexane. Therefore, despite the presence of an active hydroxy-group in the molecules of 1-octanol and electronegative chlorine and fluorine atoms in the structure of the dissolved substances, the solubility is not determined by the hydrogen bonding ability but by non-specific salvation. It is worth noting that the low solubility of the studied substances in 1-octanol is also caused by the stronger association and bigger molar volume of the alcohol molecules compared to those of hexane, which hinders the penetration of the molecules of the dissolved substance into the free cavity of the solvent.

Table 5. Thermodynamic solubility functions of the studied compounds in buffer pH 7.4, hexane and 1-octanol at 298.15 K

Compound	[a] ΔG^0_{sol} (kJ/mol)	[b] ΔH^0_{sol} (kJ/mol)	[c] ΔS^0_{sol} (J/mol·K)	ΔG^0_{sol} (kJ/mol)	ΔH^0_{sol} (kJ/mol)	ΔS^0_{sol} (J/mol·K)	ΔG^0_{sol} (kJ/mol)	ΔH^0_{sol} (kJ/mol)	ΔS^0_{sol} (J/mol·K)
	buffer pH 7.4			hexane			1-octanol		
I	32.7	13.7	-63.73	28.6	25.6	-6.7	20.3	34.7	69.9
II	32.6	25.4	-24.2	27.2	24.9	-7.7	22.5	32.3	32.9
III	29.6	22.6	-23.4	25.3	18.0	-24.5	23.0	40.5	58.7
IV	28.6	21.3	-24.6	19.1	21.8	9.1	20.1	29.1	30.2
V	30.7	31.6	2.9	18.2	24.3	13.1	21.9	36.8	50.0

[a] ΔG^0_{sol} - standard molar dissolution Gibbs energy. [b] ΔH^0_{sol} - standard molar dissolution enthalpy.

[c] ΔS^0_{sol} - standard molar dissolution entropy.

The solubility data are represented as linear functions ln$x$ *versus* 1/T, and the thermodynamic solubility parameters of the compounds in the buffer (pH 7.4), hexane and 1-octanol are calculated by equations (2 – 5) and summarized in Table 5.

In all the studied systems "dissolved substance – solvent," the solubility enthalpies are positive, which indicates the endothermicity of the solubility processes. The systems with hexane have the lowest solution enthalpy values for the studied substances. These values are higher in water solutions and the highest in the systems with 1-octanol. The negative entropy contribution makes the Gibbs energy higher and, as a result, lowers the substance solubility in the buffer and hexane. On the opposite, the positive entropy values characterizing the dissolution process of compounds **IV** and **V** in hexane increase the solubility of these substances up to the values exceeding their solubility in 1-octanol. When these compounds are dissolved in 1-octanol, their high positive enthalpy and entropy contributions produce a compensation effect on the Gibbs energy. As a result, the solubility of compounds **I** – **III** in 1-octanol is higher, while that of compounds **IV** and **V** is lower than in hexane.

Based on the obtained data about the solubility and ability of the synthesized substances to block glutamate-induced calcium ion uptake into rat cerebral cortex synaptosomes, we compared their biological activity at the concentration of 1 μM corresponding to the maximum Dimebon activity in this test. The analysis has shown that the best inhibition parameters (~ 17%) are characteristic of derivatives **III** with a methoxy-group and **IV** with a fluorine atom as the substituent. These compounds are better soluble in the buffer (pH 7.4), and compound **IV** also dissolves in hexane and 1-octanol better than the other derivatives.

3.5. Lipophilicity

According to the study of protolytic properties (Figure 5), the molecules of the compounds **I** – **V** in the buffer pH 7.4 are in the neutral form. That is why the experimentally obtained partition coefficients of the synthesized derivatives in the system 1-octanol/buffer pH 7.4 represent the ratio of the nonionized molecule concentrations, and the log*P* values characterize the compound lipophilicity. The log*P* values at the temperature of 298.15 K shown in Table 6 are within the range of 0.47 – 1.30, which indicates low lipophilicity of the investigated substances. It should be noted that the drug lipophilicity limit determined empirically is expressed by the inequality: -0.5 < log*P* < 5 (Ghose et al., 1999).

Table 6. Experimental concentrations expressed in mol/l (C_o–in 1-octanol, C_b–in buffer pH 7.4) and partition coefficients (log*P*) for the studied compound in the system 1-octanol / buffer pH 7.4 at different temperatures and pressure *p*=0.1 MP

T (K)	**I**			**II**		
	$C_o \cdot 10^4$	$C_b \cdot 10^5$	log*P*	$C_o \cdot 10^4$	$C_b \cdot 10^5$	log*P*
293.15	1.17	4.31	0.43	1.02	1.56	0.81
298.15	1.20	4.02	0.47	1.03	1.48	0.84
303.15	1.23	3.73	0.52	1.04	1.41	0.87
308.15	1.24	3.55	0.54	1.05	1.48	0.89
313.15	1.27	3.32	0.58	1.06	1.05	0.92
T (K)	**III**			**IV**		
	$C_o \cdot 10^4$	$C_b \cdot 10^5$	log*P*	$C_o \cdot 10^4$	$C_b \cdot 10^5$	log*P*
293.15	1.08	2.72	0.60	1.97	3.74	0.72
298.15	1.10	2.55	0.63	1.99	3.45	0.76
303.15	1.12	2.36	0.68	2.02	3.21	0.80
308.15	1.13	2.21	0.71	2.05	3.05	0.84
313.15	1.15	2.06	0.75	2.07	2.72	0.88
T (K)	**V**					
	$C_o \cdot 10^4$	$C_b \cdot 10^5$	log*P*			
293.15	1.43	0.76	1.27			
298.15	1.43	0.73	1.29			
303.15	1.43	0.69	1.32			
308.15	1.44	0.66	1.34			
313.15	1.44	0.64	1.35			

In order of increasing partition coefficients, the compounds are arranged in the following series: **I** < **III** < **IV** < **II** < **V**. Compound **V** with a fluorine atom as the substituent has the maximum value of the partition coefficient, whereas unsubstituted compound **I** has the lowest partition value. The obtained data are quite reasonable as the substitution of the hydrogen atom in the aromatic nucleus with halogen atoms leads, as a rule, to an increase in the compound lipophilicity (Arnott and Planey, 2012; Thomas, 2000). Compound **III** has a lower partition coefficient value than compound II because the replacement of the nonpolar methyl group with a polar methoxy-group results in molecule hydrophility growth.

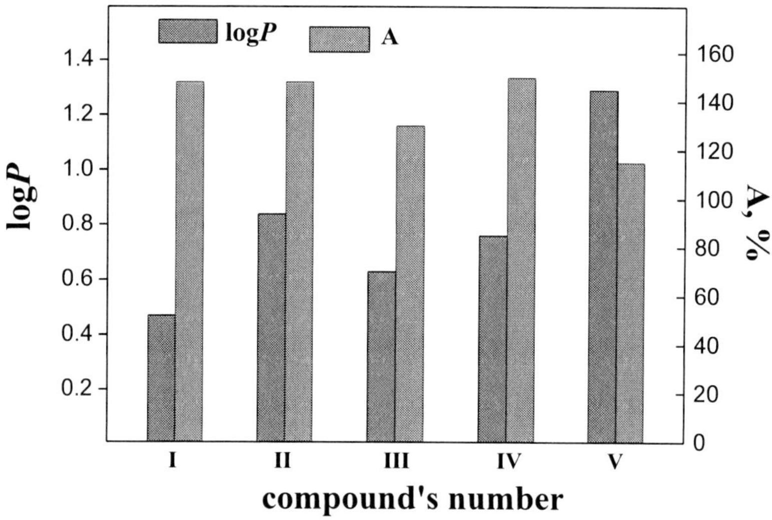

Figure 7. Histogram of partition coefficients (log*P*) of the studied compounds in the system 1-octanol/buffer pH 7.4 and amplitudes of kainic acid-induced currents (A) in Purkinje cells.

Figure 7 shows the results of investigating the biological activity and lipophilicity of the synthesized compounds.

The highest activity in this test is observed in compounds **I**, **II** and **III**, the log*P* values of which change within the range of 0.47 – 0.84. The log*P* value of derivative **III** also belongs to this range but the activity of this substance is lower. The replacement of the methyl group in compound **II** with a methoxy-one in compound **III** is accompanied by a reduction in the lipophilicity and activity. Despite the high lipophilicity of compound **V**, its biological activity is rather low. It should be noted that in the tests aimed at finding out the ability of the studied substances to inhibit calcium ion trapping in rat cortex synaptosomes, derivatives **I** and **III** had the highest degree of biological activity. And, as our experiment has shown, these substances have the lowest lipophilicity.

In terms of comparing the bioactivity of the synthesized substances, it is interesting to find correlations between the concentrations at which the compounds have the highest activity and the partition coefficient logarithms. Hansch et al. (Hansch, 1991) have derived a parabolic regression equation that can quite adequately describe the dependence between the partition coefficient and the drug concentration triggering a standard biological response. The obtained results show that with increasing lipophilicity of the compounds studied, the maximum biological activity in this test is achieved at a higher concentration of the administered substance (Figure 8).

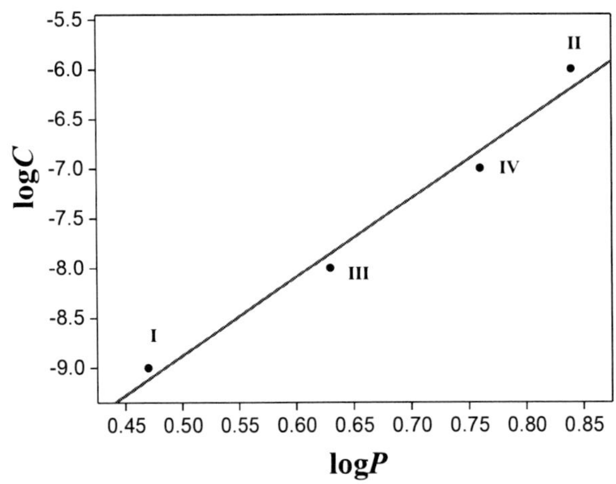

Figure 8. Correlation partition coefficients (log*P*) of the studied compounds in the system 1-octanol/buffer pH7/4 with the concentration (*C*, mol/l) of maximal potentiating action on AMPA receptors.

Compound **V** that is the least biologically active was excluded from the test. The obtained correlation log(1/*C*)=*f*(log*P*) indicates that there is a linear dependence between the logarithms of the two variables: the introduced drug dose and its lipophilicity. When the lipophilicity of the studied substances grows, the therapeutic effect can be achieved by increasing the drug dose.

For example, at the same maximum bioactivity, the content of compound I (log*P*=0.47) in the solution bathing the neuron is three orders of magnitude lower than that of compound **II** (log*P*=0.84). It is worth noting that in most cases, lipophilicity growth makes compounds more active as it becomes easier for them to permeate through the membranes and to achieve the biological target (Kerns and Di, 2008). However, if lipophilicity is not the main factor determining the biological action of a drug, but only creates additional conditions for this, the situation can change. For example, the activity of penicillins increases slightly if their hydrophilicity grows (Albert, 2012). The increase in biological activity with the lipophilicity decrease found in this work is most probably caused by higher energy of interaction of the studied hydrophilic compounds with the receptor.

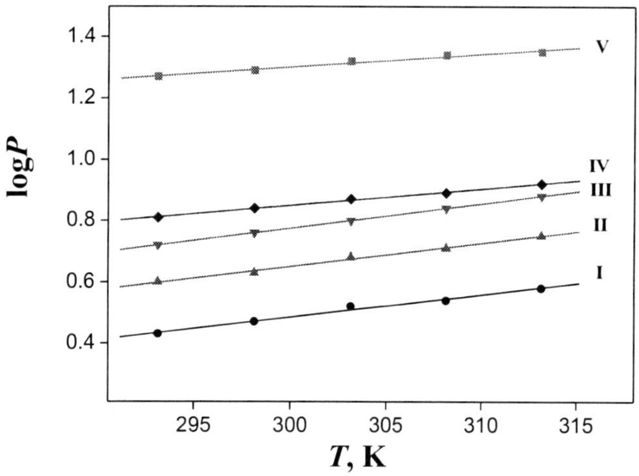

Figure 9. Temperature dependences of partition coefficients (log*P*) for the studied compounds in the system 1-octanol/buffer pH 7.4.

Table 6 contains log*P* values of the studied compounds measured in the temperature range of 293.15 – 313.15 K. For all the substances, the log*P* temperature dependence graphs are linear, with the correlation coefficient R>0.9978 (Figure 9).

The temperature growth results in a partition coefficient increase, which indicates that the content of the substance rises in the octanol phase imitating the membrane lipid layer. Based on the temperature dependences of the partition coefficients, we have calculated the thermodynamic functions of compound transfer from the aqueous phase to the organic one by equations (2 - 4). The results are summarized in Table 7.

The fact that the changes in the Gibbs energy are negative means that the compound transfer from the buffer solution to the organic medium is quite favourable. We have determined that the H_{tr} values of the studied substances are positive. The endothermic character of the transfer indicates that the energy of the dissolved substance interactions with the solvent in the octanol phase is lower than the energy required for breaking the intermolecular bonds in the aqueous medium. Such characteristics of the intermolecular interactions are associated with the hydrophobic hydration of phenyl and alkyl molecule fragments of the studied compounds. The changes in the transfer entropy ΔS^0_{tr} r are also positive and, consequently, the compound transfer makes the system less ordered. It is worth noting that the partition process of more active compounds **I** – **IV** is characterized by higher endothermicity than that of compound **V** that is less active in this test.

Table 7. Thermodynamics functions of transfer in the system 1-octanol/buffer pH 7.4 for the studied compounds at 298.15 K

Compound	ΔG^o_{tr} (kJ·mol^{-1})	ΔH^0_{tr} (kJ·mol^{-1})	TΔS^0_{tr} (kJ·mol^{-1})	ΔS^0_{tr} (J·mol^{-1}·K^{-1})
I	-8.1 ± 0.1	12.9 ± 0.4	21.0 ± 1.0	70.4 ± 1.5
II	-10.2 ± 0.2	8.4 ± 0.1	18.6 ± 0.8	62.4 ± 1.9
III	-9.0 ± 0.2	13.0 ± 0.2	22.0 ± 0.9	73.8 ± 2.5
IV	-9.7 ± 0.2	14.2 ± 0.3	23.9 ± 1.1	80.3 ± 2.5
V	-12.8 ± 0.3	6.9 ± 0.1	19.7 ± 0.8	65.9 ± 2.2

CONCLUSION

In this study, hydrogenated pyrido[4,3-b]indoles as potential drug compounds with neurocorrecting activity have been synthesized. The obtained compounds differ in the chemical nature of the substituent (H-, CH_3-, CH_3O-, F- Cl-) introduced into position 8 of the carboline fragment of the molecule. The biological tests have shown that all the studied compounds can modulate the glutamate-dependent uptake of calcium ions in rat cerebral cortex synaptosomes. The compound with a methoxy-substituent has been found to have the highest inhibiting activity within a broad concentration range, which allows us to recommend it as the object of further research. The influence of the compound structure on the functional activity of AMPA-dependent ionic channels was studied by the physiological patch-clamp method on single isolated Purkinje neurons excreted from rat brain cerebellum. Adding the studied substances to the solution bathing the neuron led to concentration-dependent changes in the current amplitude compared to the reference substance. The most active compounds in this test (~ 150% of the reference substance value) were the unsubstituted derivative and the compounds with a methyl group and a fluorine atom. The bioactivity of the compound with a methoxy-group was a little lower (~ 130% of the reference substance value), while the fluorine-derivative had a weak effect on the kainate-activated currents. The maximum positive modulation of the AMPA receptors was achieved if the concentrations of the introduced substances were in the range of 10^{-9}-10^{-6} mol/l.

The shake-flask method was used to measure the solubility of the studied compounds in three pharmaceutically relevant media: a buffer solution (pH 7.4), 1-octanol and hexane. It has been identified that all the derivatives in these solvents have a very low solubility not exceeding the value of 8 10^{-4} mole fractions. And the compounds with a methoxy-group and a fluorine atom as substituents are best soluble in the buffer solution (pH 7.4) modeling the blood plasma medium.

In order to find the reasons for the poor solubility of the studied compounds, their thermophysical and protolytic properties were analyzed. By using the DCS method it has been found that these substances melt at the temperature above 230°C. This indirectly proves that the crystal lattice energy of the compounds is high and is one of the factors lowering the solubility. Besides, the obtained experimental and calculation data have shown that in the studied solutions the molecules are in the neutral form and interact with the solvent only through non-specific salvation, which also increases the solubility. Based on the temperature dependences of solubility, the thermodynamic functions of substance dissolution in the solvents used have been calculated. In the studied systems "dissolved substance – solvent," the dissolution processes are endothermic. In all the studied substances, when the substituent is replaced, the solution enthalpy increases in the following order: hexane, buffer (pH 7.4), 1-octanol.

The partition coefficients of the synthesized derivatives in the system 1-octanol/buffer pH 7.4 in the temperature range of 293.15 – 313.15 K have been measured. We have found that the logP values of the studied compounds are in the range of 0.47 - 1.30, which indicates their low lipophilicity. And the most biologically active substances are the unsubstituted derivative, and the compounds with a methyl group and a fluorine atom as substituents. The obtained results show that when the lipophilicity of the studied compounds increases, the maximum biological activity in this test is achieved at a higher concentration of the administered substance. It has been established that the increase in the biological activity with a lipophilicity decrease is caused by higher interaction energy of the hydrophilic molecular forms of the studied compounds with the receptor. Based on the temperature dependences of the partition coefficients, we have calculated the thermodynamic functions of compound transfer from the buffer solution to 1-octanol. For all the studied compounds, the transfer process is endothermic and, consequently, the interaction energy of the dissolved substances with the solvent in the octanol phase is lower than the energy required for breaking the intermolecular bonds in an aqueous medium.

ACKNOWLEDGMENTS

This work was supported by the grant of Russian Foundation for Basic Research No. 18-43-370016.

References

Albert, A. (2007). *Selective Toxicity:* The physico-chemical basis of therapy, 9th edition, Chapman and Hall, London.

Archer, D. G., Steffen, P., Rudtsch, J. (2003). Enthalpy of fusion of indium: a certified reference material for differential scanning calorimetry. *J. Chem. Eng. Data.* 48: 1157–1163.

Arnott, J. A., Planey, S. L. (2012) The influence of lipophilicity in drug discovery and design, *Cur. Top. Med. Chem.* 13: 1317–1326.

Asirvatham, S., Dhokchawle, B. V., Tauro, S. J. (2016). Quantitative structure activity relationships studies of non-steroidal anti-inflammatory drugs: A review. *Arab. J. Chem.* https://doi.org/10.1016/j.arabjc.2016.03.002

Avdeef, A. (2007). Solubility of sparingly-soluble ionizable drugs. *Adv. Drug Deliv. Rev.* 59: 568–590.

Baka, E., Comer J. E. A., Takacs-Novak, E. (2008). Study of equilibrium solubility measurement by saturation shake-flask method using hydrochlorothiazide as model compound. *J. Pharm. Biomed. Anal.* 46: 335–341.

Blokhina S., Sharapova A., Ol'khovich M. et al. (2018) New derivatives of hydrogenated pyrido[4,3-b]indoles as potential neuroprotectors: Synthesis, biological testing and solubility in pharmaceutically relevant solvents. *Saudi Pharm. J.* https://doi.org//10.1016/j.jsps.2018.04.003.

Caron G., Reymond F., Testa B. (1999) Combined molecular lipophilicity descriptors and their role in understanding intramolecular effects. *Pharm. Sci. Technol. Today* 2:327–335.

Chang R. C. C. (2011) *Advanced Understanding of Neurodegenerative Diseases.* Rijeka, InTech.

Danishuddin, Khan, A. U. (2016). Descriptors and their selection methods in QSAR analysis: paradigm for drug design. *Drug Discov.* 21: 1291-1302.

Dondas, H. A., Hempshall, A., Narramore, S., Kilner, C., Fishwick, C. W. G., Grigg, R. (2016). γ-Carboline AC190 analogues via palladium catalysed allene insertion stereo and regioselective 3- and 5-component cascades. *Tetrahedron*. 72: 1316-1329.

Doody, R. S., Gavrilova, S. I., Sano, M., Thomas, R. G., Aisen, P. S., Bachurin, S. O., Seely, L., Hung, D. (2008). Effect of dimebon on cognition, activities of daily living, behaviour, and global function in patients with mild-to-moderate Alzheimer's disease: a randomised, double-blind, placebo-controlled study. *Lancet*. 372: 207–215.

Du-Cuny, L., Huwyler, J., Wiese, M., Kansy, M. (2008). Computational aqueous solubility prediction for drug-like compounds in congeneric series, *Eur. J. Med. Chem.* 43: 501-512.

Eckert G. P., Renner, K., Eckert, S. H., Eckmann, J., Hagl, S., Abdel-Kader, R. M., Kurz, C., Leuner, K., Muller, W. E. (2012). Mitochondrial dysfunction - a pharmacological target in Alzheimer's disease. *Mol. Neurobiol.* 46: 136-50.

Faller, B., Erlt, P. (2007). Computational approaches to determine drug solubility. *Adv. Drug Deliv. Rev.* 59: 533–545.

Farlow, M. R. (2004). NMDA receptor antagonists: a new therapeutic approach for Alzheimer's disease. *Geriatrics*. 59: 22-27.

Ghose A. K., Viswanadhan V. N., Wendoloski J. J. (1999) A knowledge-based approach in designing combinatorial or medicinal chemistry libraries for drug discovery. 1. A qualitative and quantitative characterization of known drug databases. *J. Comb. Chem*. 1: 55–68.

Gylys, K. H., Fein, J. A., Cole, G.M. (2000). Quantitative characterization of crude synaptosomal fraction (P-2) components by flow cytometry. *J. Neurosci. Res.* 15: 186-192.

Hansch C. (1971) Quantitative structure-activity relationship in drug design, in: Ariens E. J. (Ed.), *Drug Design*. New York, Academic Press; p 271−342.

Hajos, F. (1975). An improved method for the preparation of synaptosomal fractions in high purity. *Brain Res.* 93: 485-489.

Hung, D. T., Protter, A. A., Jain, R. P., Dugar, S., Chakravarty, S. (2015). *Tetracyclic compounds*. Patent US8999978 B2.

Ikonomidou, C., Turski, L. (2002). Why did NMDA receptor antagonists fail clinical trials for stroke and traumatic brain injury? *Lancet Neurol.* 1: 383-386.

Ivachtchenko, A. V., Frolov, E. B., Mitkin, O. D., Tkachenko, S. E., Khvat A. V. (2010). Synthesis and biological activity of 5-styryl and 5-phenethyl-substituted 2,3,4,5-tetrahydro-1H- pyrido[4,3-b]indoles. *Bioorg. Med. Chem. Let.* 20: 78-82.

Kaneda M, Nakamura H, Akaike N. (1988). Mechanical and enzymatic isolation of mammalian CNS neurons. *Neurosci. Res.* 5: 299-315.

Khachaturian, Z. S. (1994). Calcium hypothesis of Alzheimer's disease and brain aging, *Ann. N. Y. Acad. Sci.*, 747: 1–11.

Kerns, E., Di, L. (2008). *Drug like Properties: Concepts, Structure Design and Methods*, Academic Press, New York.

Khritankova I. V., Kukharskiy M. S., Lytkina O. A., et al. (2012). Dimebon activates autophagosome components in human neuroblastoma SH-SY5Y cells. *Dokl. Biochem. Biophys.* 446: 251−253.

Lynch G. (2006). Glutamate-based therapeutic approaches: ampakine, *Curr. Opin. Pharmacol.* 6: 82–88.

Mannhold R., Poda G.I., Ostermann C., Tetko I. V. (2009). Calculation of molecular lipophilicity: state-of-the-art and comparison of logP methods on more than 96.000 compounds. *J. Pharm Sci.* 98:861–893.

Mattson, M. P., Cheng, B., Davis, D., Bryant, K., Lieberburg, I., Rydel, R. E. (1992). β-Amyloid peptides destabilize calcium homeostasis and render human cortical neurons vulnerable to excitotoxicity, *J. Neurosci.* 12: 376–389.

Molleman A. (2003). *Patch Clamping: An Introductory Guide to Patch Clamp Electrophysiology.* Chichester, John Wiley & Sons.

Parsons, C. G., Danysz, W., Quack, G. (1998). Glutamate in CNS Disorders as a Target for Drug Development: An Update. *Drug News Perspect.* 11: 523-579.

Po, H. N., Senozan, N. M. (2001). Henderson-Hasselbalch Equation: Its History and Limitations. *J. Chem. Educ.* 78: 1499-1503.

Rekker, F. R., Mannhold, R. (1992). *Calculation of drug lipophilicity: the hydrophobic fragmental constant approach.* VCH, Weinheim.

Sangster J. (1997). *Octanol-water partition coeffcients: fundamentals and physical chemistry.* Chichester, John Wiley & Sons.

Shaffer C. L., Hurst R. S., Osgood S. M., et al. (2013). Positive allosteric modulation of AMPA receptors from efficacy to toxicity: the interspecies exposure-response continuum of the novel potentiator PF-4778574. *J. Pharmacol. Exp. Ther.* 347: 212−224.

Steele, J. W., Gandy, S. (2013). Latrepirdine (Dimebon®), a potential Alzheimer therapeutic, regulates autophagy and neuropathology in an Alzheimer mouse model. *Autophagy*. 9: 617–618.

Stefek, M., Tsantili-Kakoulidou, A., Milackova, I., Juskova, M., Triantos, N. (2011). (2-Benzyl-2,3,4,5-tetrahydro-1H-pyrido[4,3-b]indol-8-yl)-acetic acid: An aldose reductase inhibitor and antioxidant of zwitterionic nature. *Bioorg. Med. Chem.* 19: 7181-7185.

Stephen, H., Stephen, T., (1963). *Solubilities of inorganic and organic compounds.* Oxford Pergamon, London.

Thomas G. (2000) *Medicinal Chemistry an Introduction.* West Sussex, John Wiley & Sons.

Van de Waterbeemd, H., Rose, S. (2008). Quantitative approaches to structure – activity relationships, in: Wermuth C. (Ed.), *The practice of Medicinal Chemistry,* third edition, pp. 491-513.

Wang T., Wu M. B., Lin J. P., et al. (2015) Quantitative structure-activity relationship: promising advances in drug discovery platforms. *Expert Opin. Drug Discov.* 10:1283−1300.

In: Indole: Synthesis, Functions and Reactions ISBN: 978-1-53614-777-3
Editor: Bridget Pratt

Chapter 2

MUSHROOMS AS A SOURCE OF BIOLOGICAL ACTIVE INDOLE COMPOUNDS

Bożena Muszyńska**, Katarzyna Kała and Katarzyna Sułkowska-Ziaja
Department of Pharmaceutical Botany, Faculty of Pharmacy,
Jagiellonian University Medical College, Kraków, Poland

ABSTRACT

Mushrooms and their mycelia from *in vitro* cultures are increasingly common study subjects due to their health-promoting potential. The most frequently demonstrated activity of these mushrooms include antioxidative, anticancer, anti-inflammatory, immunostimulatory and anti-depressive effect. This activity stems from numerous mushroom metabolites, bioelements and physiologically active substances, including indole compounds. Indole compounds present in mushrooms can be divided into tryptamine derivatives with hallucinogenic properties and bioactive indole derivatives, without hallucinogenic properties. In numerous species of edible mushrooms, including species living on wood, both the fruiting bodies as well as mycelia from *in vitro* cultures were found

* Corresponding Author Email: muchon@poczta.fm.

to contain non-hallucinogenic indole derivatives: 5-hydroxy-L-tryptophan, serotonin, L-tryptophan, tryptamine, melatonin, 5-methyl-D,L-tryptophan, indole-3-acetic acid, and indole. Moreover, the influence of substrate enriching for mycelial cultures with specific substances, including element salts on the profile of indole compounds and their total content has also been confirmed. These studies have enabled standardization of the conditions of *in vitro* culture conducting in laboratory, so as to obtain the highest possible contents of indole compounds in their biomass. Moreover, the influence of heat processing, analogous to cooking, has also been examined (as a method of preparing meals based on mushrooms) on the content of indole compounds. It has been demonstrated that heat processing results in approximately two-fold decrease of the content of indole compounds, but mushrooms remain their good source. It has also been demonstrated that fruiting bodies of selected edible mushroom species and their *in vitro* cultures not only synthesize indole derivatives and accumulate them, but they also release these substances to the solution of artificial gastric juice and artificial intestinal juice under *in vitro* conditions. Thus, they may constitute good source of indole compounds for human organism. The obtained highest degree of release of indole compounds from *in vitro* cultures in conditions stimulating natural digestive processes is particularly significant, as it enables development of commercial cultures accumulating these physiologically significant compounds, and in the future obtaining of enriched fruiting bodies, which may be used to produce natural diet supplement. Additionally, it should be emphasized that the presence of indole compound such as L-tryptophan and 5-hydroxy-L-tryptophan in fruiting bodies of edible mushrooms species additionally enhances their dietary values, which is of particular significance in vegetarian diet, where mushrooms rich in indole derivatives may constitute an alternative to red meat. In summary, indole compounds are successfully determined in mushrooms that demonstrate a multi-directional therapeutic and dietary activity.

Keywords: edible mushrooms, indole compounds, pro-health potential, dietary properties, antidepressant activity

INTRODUCTION

The search for natural raw materials being the source of substances with therapeutic importance is highly popular nowadays [1-3]. Raw materials appreciated for centuries are being used. This group undoubtedly includes mushroom fruiting bodies, of both edible and inedible species that exhibit certain therapeutic properties. A group of therapeutic mushrooms has already been distinguished due to their numerous properties. Basidiomycota mushrooms, that is possessing spores numbering 2–4 attached to the so-called basidia, are dominant among the distinguished species [4, 5]. Mushrooms are highly popular not only due to their taste and aroma but also due to a range of activities presented. Here, anti-cancer, neuroprotective, anti-depressive, anti-oxidative, immunostimulatory, anti-atherosclerotic, anti-inflammatory, anti-allergic, antimicrobial and even antidiabetic effect can be mentioned. A significant dietary aspect of edible mushrooms may not be omitted. Despite the high water content in fresh fruiting bodies (from 70 to 95%), dried mushrooms contain from 0.5 to 3.5% fat, up to approx. 5% carbohydrates and from 3% to 25% of assimilable proteins [1, 6-10].

The range of therapeutic activities exhibited by mushrooms are produced by physiologically active components, including the best understood fungal metabolite group, that is polysaccharides, substances with protein character (glycoproteins, lectins or fungal immunomodulating proteins), terpenoids, phenolic compounds, lipids (incl. ergosterol and its derivatives) or indole compounds [1, 3, 9-11].

In addition, due to the periodic occurrence of fruiting bodies of numerous mushroom species, the biotechnological aspect and obtaining of mycelial cultures in laboratory conditions is highly significant. Mycelial cultures may constitute a source of fungal physiologically active compounds, including indole compounds. Moreover, standardization of mycelial cultures conduct, including temperature, lighting conditions or specific additives to culture media may have a positive impact on the total contents of important fungal metabolites such as indole compounds, in the obtained biomass. Thanks to the development of the branch of biotechnology, the last four decades have been the period of the most intensive research on substances isolated from mushrooms and their *in vitro* cultures [12-14].

Indole compounds determined in fruiting bodies and *in vitro* cultures of mushrooms constitute a very important group of substances. These compounds contain coupled benzene and pyrrole rings [15]. Among indole compounds of fungal origin, health-promoting, toxic (ergot alkaloids) or hallucinogenic substances to the human organism can be found. An example of a hallucinogenic, indole compound originating from mushrooms are lysergic acid derivatives, which include LSD. It is among the most active psychoactive compounds discovered so far-typically its dose ranges between 100 and 500 micrograms. It produces visual, auditory and tactile hallucinations. Moreover, indole derivatives can be found among commonly used therapeutic substances. These include indomethacin-a non-steroidal anti-inflammatory drug, or ziprasidone and sertindole, used as antipsychotic drugs [15-18].

Non-hallucinogenic indole derivatives (determined in mushrooms and mycelial cultures) with therapeutic importance include, among others serotonin, melatonin, L-tryptophan, 5-hydroxy-L-tryptophan and tryptamine. These compounds fulfill the role of neurotransmitters or their precursors. They impact the daily cycle, regulating appetite, mood, and bodily temperature. They are responsible for coagulation processes, immune response, and cellular regeneration. Moreover, they demonstrate antioxidative (thus reducing lipid peroxidation), anticancer, cardioprotective and analgesic properties [17, 19, 20].

Biogenesis of Indole Compounds

The basic structural element of indole compounds is the indole core or reduced dihydroindole core (indoline) core. Biosynthetically, this skeleton is formed out of tryptophan or product of its decarboxylation of tryptamine. Tryptophan is formed of shikimic acid in two ways, in which anthranilic acid plays a significant role. The fragment without tryptophan may be of a monoterpene origin, and the compounds necessary for synthesis include mevalonic acid, geranil or nerol, loganin, and secologanin [21]. The metabolic pathway of tryptophan is presented in Figure 1.

Simple derivatives of tryptamine are classified among amines or protoalkaloids, harmane or β-carboline derivatives may also occur.

Compounds with a more advanced structure possess, apart from the main skeleton, a fragment made of 9–10 carbons or a dimeric structure [21].

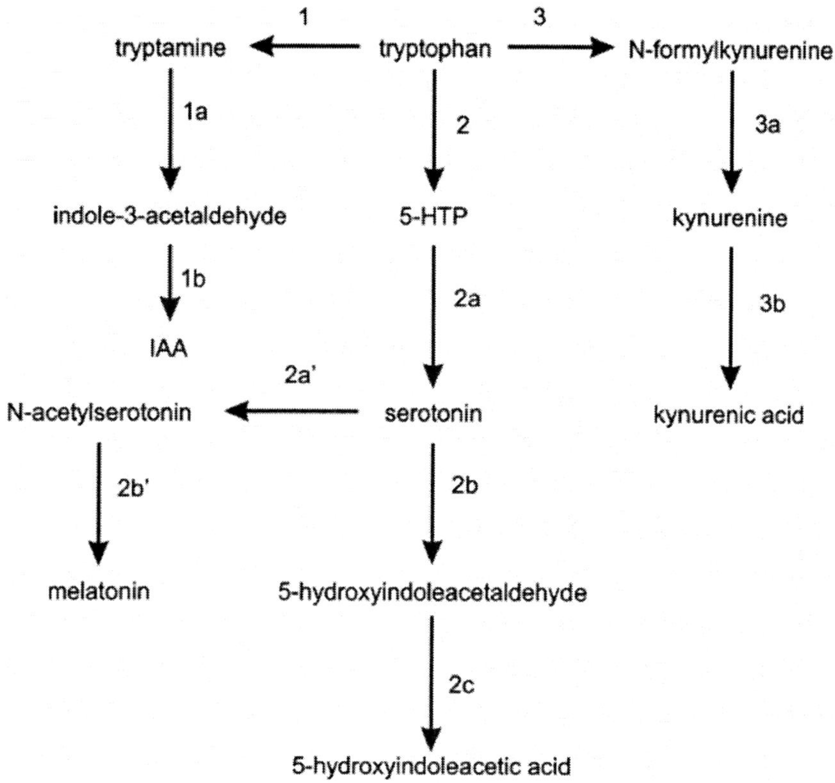

Figure 1. Tryptophan metabolic pathway: 1. L-tryptophan decarboxylase, 1a. tryptamine deaminase, 1b. aldehyde dehydrogenase, 2. tryptophan hydroxylase, 2a. aromatic amino acid decarboxylase, 2b. monoamine oxidase, 2c. aldehyde dehydrogenase, 2a'. serotonin N-acetylase, 2b'. 5-hydroxyindole O-methyltransferase, 3. formylkynureninase, 3a. formamidase, 3b. kynurenine transaminase, IAA–indole-3-acetic acid, 5-HTP–5-hydroxytryptophan.

Biological Characteristics of Indole Compounds

Macromycetes synthesize numerous compounds with the structure of indole. Those compounds present in mushrooms can be divided into tryptamine derivatives with hallucinogenic properties and bioactive indole derivatives, without hallucinogenic properties [22, 23].

The main substances with hallucinogenic properties are psilocybin (4-phosphonyloxy-N,N-dimethyltryptamine) and its accompanying second alkaloid, psilocin (4-hydroxy-N,N-dimethyltryptamine). Apart from these compounds, mushrooms have been found to also contain baeocystin (4-phosphoryloxy-N-methyltryptamine) which is a psilocybin derivative, differing from it only by the lack of methyl substituent at the nitrogen in its side chain and norbaecestin (4-phosphoryloxytryptamine) typically present at trace amounts [16, 18, 23, 24].

Table 1. L-tryptophan content in example foods

Product	L-tryptophan content (g/100g of product)	Product	L-tryptophan content (g/100g of product)
Pork meat	0.32	Sesame (seeds)	0.37
Poultry meat	0.30	Cocoa	0.29
Salmon	0.25	Walnuts, hazelnuts	0.19
Cheese	0.35	Spinach	0.1
Milk	0.04	Banana	0.01

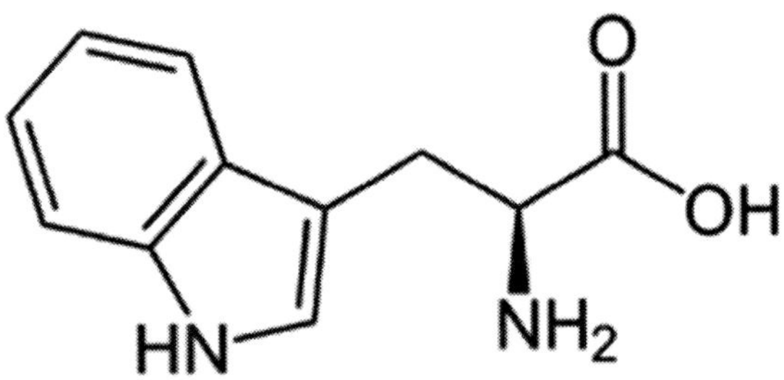

Figure 2. L-Tryptophan.

Indole is the precursor of numerous substances found in human organism, such as L-tryptophan, serotonin, and melatonin. Indole is produced by various microorganisms, e.g., from the family *Enterobacteriaceae*. In very high dilution this compound possesses a jasmine odor-as it is one of the compounds of natural jasmine oil. It is used in the perfume industry to produce synthetic jasmine oil, which is considerably less expensive from its natural equivalent. In higher concentrations, its odor resembles concentrated manure odor (its component, the so-called animal indican) [25-27].

Indole derivatives such as L-tryptophan, which is an essential amino acid for humans, must be provided to the organism with food, in order to ensure its normal functioning. Therapeutic mushrooms, along with other foods, constitute their good source (Table 1) [5, 28].

In the human central nervous system tryptophan is transformed into serotonin and melatonin. L-Tryptophan undergoes three major transformations: decarboxylation leading to the formation of tryptamine, hydroxylation with transformation into 5-hydroxytryptophan, from which serotonin is formed and indole ring disruption, leading to obtaining kynurenine. Moreover, it is a niacin precursor. Its sleep-inducing and anti-depressive effect have been known for a long time (it penetrates the blood-brain barrier and it is metabolized to serotonin) [25-27]. L-tryptophan (Figure 2) is sometimes combined with other anti-depressive drugs in a depression therapy. Moreover, it constitutes an ingredient of dietary supplements intended for use in the case of depressive conditions, stressful situations, and sleep disorders.

5-Hydroxytryptophan, also common in mushrooms, is the direct serotonin precursor. Consumed with food, it easily penetrates the blood-brain barrier in the central nervous system, where it is transformed to serotonin (serotonin does not penetrate the blood-brain barrier) [29, 30]. In certain countries (United Kingdom, USA, Canada) it is used as a diet supplement with anti-depressive, appetite-reducing and sleep-inducing effect. To this end, it is extracted from the seeds of an African plant *Griffonia simplicifolia* [30]. Absorption of 5-hydroxytryptophan does not require the presence of transport molecule. Moreover, the absorption is not affected by the presence of other amino acids, thus 5-hydroxytryptophan may be consumed with food, without resulting in undesirable interactions. This compound is characterized by good absorption after oral administration, with about 70% of the dose penetrating to the blood stream. It has been demonstrated that administration of 5-hydroxytryptophan in therapy can be efficient in treating such diseases as depression, fibromyalgia, chronic headaches and insomnia [30, 31].

Serotonin (5-hydroxytryptamine) (Figure 3) was isolated in 1937 from enterochromatophilic cells present in the gastrointestinal tract.

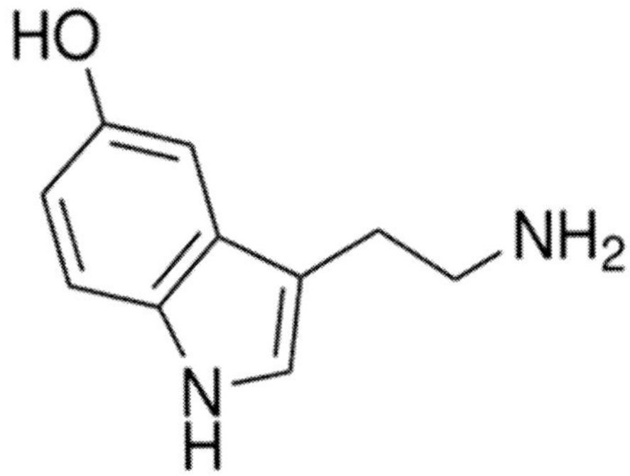

Figure 3. Serotonin.

Its discoverer was searching for a substance responsible for smooth muscles' spasm. It was named 'enteramine'. In 1948, Maurice Rapport and Irvine Page isolated and described 5-hydroxytryptamine, when seeking a substance responsible for vasoconstriction, and then it was named serotonin from the Latin word *serum* and Greek *tonic.* In 1952 it was demonstrated that serotonin and enteramine are the same compounds. Serotonin, a substance with a multi-directional pharmacological activity, performing i.a. the role of a neurotransmitter in the central nervous system regulates the daily cycle with melatonin. Moreover, it is produced not only in the brain but also by enterochromatophilic cells of the intestines, it participates in the spasm of smooth muscles, regulates bowel function, affects regulation of blood pressure, participates in the blood coagulation process and has an antioxidative effect [26]. Serotonin, generated endogenously in the brain, has a very significant role in the regulation of sleep, mood, anxiety, aggression, bodily temperature, course of the maturation, regeneration and inhibition of process of cell ageing, thus contributing to the general strengthening of the immune system of the organism [29, 30]. It produces bronchospasm in patients with asthma. The wide range of activities regulated by serotonin explains the existence of 7 types of serotonin receptors (5-HT), and several subtypes within each receptor. Numerous drug classes act via serotonergic receptors, by being agonists or antagonists to these receptors, or through influence on the concentration of serotonin. Such drugs primarily include antidepressant agents, anxiolytics, antiemetic and antimigraine drugs. According to the conducted research, this compound is also a potential drug for the treatment of Alzheimer's disease [31-33]. Serotonin is applied in the treatment of migraines and depression at doses from 100 mg to 600 mg. *In vitro* tests have demonstrated that serotonin and melatonin, as well as other indole derivatives (N-acetylserotonin, 6-methoxytryptamine), reduce lipid peroxidation in a manner that depends on the dose [29-31, 33].

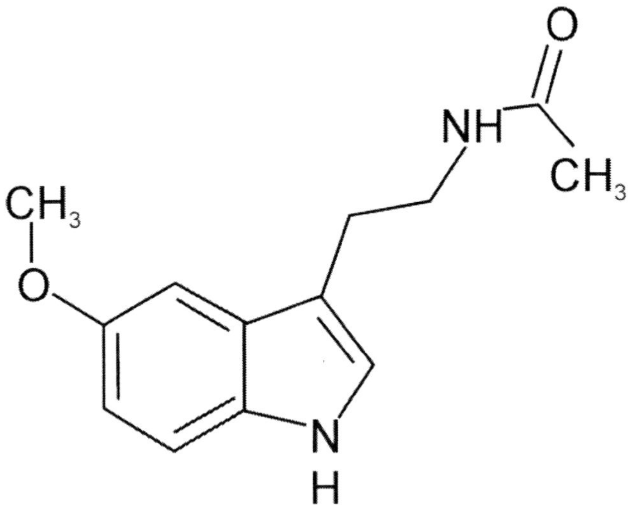

Figure 4. Melatonin.

Kynurenine is an amino acid, a metabolite of L-tryptophan used in niacin production. Kynurenine is present in the blood and in various human organs. As many as 40% of this metabolite is synthesized by the glia cells in the central nervous system, the remaining 60% originates from peripheral tissues ant it is accumulated in the plasma [25, 32]. Kynurenic acid is responsible for the adaptational processes in the human organism. This compound has been found in all mammalian organs; its highest concentration is found in the liver cells, while the lowest-in the brain. This produces dose-dependent hypotonia, and thus acts neuroprotective in the states of brain hypoxia. Moreover, kynurenic acid prevents the formation of gastric and duodenal ulcers [32]. Other indole compounds most frequently occurring in Basidiomycota are melatonin and tryptamine [34, 35].

Melatonin (N-acetyl-5-methoxytryptamine) (Figure 4) is primarily produced in pineal gland pinealocytes, as well as in the eye retina and in the gastrointestinal tract.

L-tryptophan is the precursor in this synthesis, which is subject to transformation to 5-hydroxytryptophan and then to 5-hydroxytryptamine. Serotonin is transformed to melatonin in a two-stage process. Melatonin production is inhibited by light. In darkness both production and secretion of melatonin are intensified. The pineal gland cells capture 5-hydroxytryptamine from the blood and metabolize it to melatonin and derivatives. This compound participates in the regulation of sleep, mood and reproduction-it coordinates the operation of the biological clock controlling the circadian rhythm and has anti-ageing effect, regenerating processes of cellular regeneration. Moreover, it has anticancer effect. The biological effects are caused by the activation of melatonin receptors. Melatonin is available as a medicinal product, intended for supporting use in sleep disorders associated with change of time zones or shift work [36-38]. Moreover, melatonin is a strong antioxidant, delaying ageing processes neutralizing free radicals, such as the hydroxyl radical. Thus, melatonin can reduce damage caused by certain types of the Parkinson's disease. Moreover, research has proven its efficiency in treating Alzheimer's disease. Melatonin has neuroprotective role in neurodegenerative diseases [35, 39]. Moreover, immunomodulatory and anti-inflammatory properties of melatonin have been demonstrated. Unfortunately, the level of melatonin decreases after 35 year, and considerably after 40 year of age, resulting in rapid human organism ageing and increased risk of neoplastic diseases. Serotonin, melatonin and their derivatives are also compounds with anti-inflammatory and analgesic effects [36].

Tryptamine (3-(2-aminoethyl)-indole) (Figure 5) is a biogenic amine.

It occurs in living organisms and it is synthesized from tryptophan. Its presence has also been discovered in the human brain and it is believed that it may fulfill a role of neuromodulator or neurotransmitter and that tryptamine receptors exist. Tryptamine constitutes the basis for a group of compounds known under a common name of tryptamine. This group contains numerous biologically active compounds. The best-known derivatives of this group include serotonin and melatonin. Certain tryptamines are used in medicine (e.g., sumatriptan and its derivatives) [18, 40-43]. Tryptamine is a metabolite, in the case of which interaction with drugs must be considered. It may interact with MAO inhibitor drugs and may result in intoxication of persons using these substances [27].

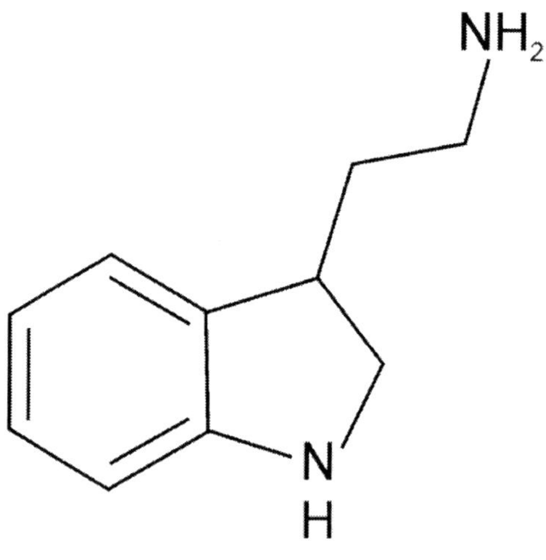

Figure 5. Tryptamine.

Other indole derivatives determined in fruiting bodies of edible mushrooms in the currently presented studies are plant auxins. Indole-3-acetic acid is a plant hormone (phytochrome) of the auxin group. This compound regulates the growth and cellular divisions. It constitutes one of the most significant auxins of natural origin. Another compound: indole-3-acetonitrile is a phytohormone of the same group and it constitutes a derivative of the indole-3-acetic acid [21].

Presence of Indole Compounds in Fruiting Bodies and Mycelial Cultures of Basidiomycota

Indole compounds present in mushrooms can be divided into tryptamine derivatives with hallucinogenic properties and bioactive indole derivatives, without hallucinogenic properties [22, 23, 44]. Numerous tryptamines with psychotropic effect have been discovered in the genus *Psilocybe* [45].

Bioactive indole derivatives with both non-hallucinogenic and psychotropic nature have been determined for *Amanita* representatives [21]. Based on the spectral analysis in 7 *Amanita* species, 12 indole derivatives have been determined: e.g., tryptamine, dimethylaminetryptamine, methoxydimethyltryptamine, L-tryptophan, N-methylserotonin, 3-indole acetonitrile. *Amanita citrina* turned out to be the richest source of these compounds.

In *Tricholoma scioides* and *Tricholoma virgatum* three indole derivatives were isolated: 2,4-dimethylindole, 4-hydroxymethyl-2-methylindole, 4-methoxymethyl-2-methylindole. Indoleacetic acid (IAA) was discovered in 16 species of mycorrhizal mushrooms [22, 44, 46-49].

Analyses of several dozens of species of wild growing mushrooms of the Ascomycota and Basidiomycota divisions, e.g., *Aleuria aurantia* Pers. Fr., *Cortinarius varius* Fr., *Cortinarius collinitus* Fr., *Gymnopilus penetrans* (Fr.) Murr., *Gymnopilus spectabilis* (Fr.) Smith., *Mycena pura* (Pers.) Quel., *Mycena rosea* (Bull.) Gromberg., *Paneolus spinctrinus* Fr. Quel., *Russula ochroleuca* (Pers.) Fr. have demonstrated the presence of numerous non-hallucinogenic indole derivatives [5, 50].

Another mushroom group subjected to analysis of indole compound content were fruiting bodies of edible mushrooms collected from the natural habitat, including *Boletus edulis* Bull. (penny bun), *Cantharellus cibarius* Fr. (golden chanterelle), *Lactarius deliciosus* Fr. (saffron milk cap), *Leccinum scabrum* Bull. Gray (a rough stemmed bolete), *Leccinum rufum*, *Macrolepiota procera* (Scop.) Singer (parasol mushroom), *Suillus bovinus* (L.: Fr.) O. Kuntze (Jersey cow mushroom), *Suillus luteus* (L.: Fries) Gray (sticky bun). Research furthermore covered fruiting bodies of edible mushrooms originating from commercial cultures: *Agaricus bisporus* (J.E. Lange) Imbach (common mushroom), *Auricularia polytricha* (Mont.) Sacc (cloud ear fungus), *Lentinula edodes* (Berk.) Pegler (shiitake) and *Pleurotus ostreatus* (Jacq. ex Fr.) P. Kumm. (oyster mushroom). Based on the obtained results it has been demonstrated that typically edible species contain significant non-hallucinogenic indole compounds, which are neurotransmitters or their precursors, such as: L-tryptophan, 5-hydroxy-L-tryptophan, 5-methyl-L-tryptophan, serotonin, melatonin, tryptamine, 5-methyltryptamine, indo, indoloacetonitryl and indoleacetic acid, indole acetonitrile and indole (Table 2 a., Table 2 b.) [5, 50].

The presence of indole compounds has also been studied in the fruiting bodies of the edible species *Armillaria mellea* (Vahl) P. Kumm. (honey fungus). Apart from tryptamine (2.74 mg/100 g d.w.) and serotonin (2.21 mg/100 g d.w.), their fruiting bodies were found to contain considerable concentrations of L-tryptophan (4.47 mg/100 g d.w.) [51].

Table 2 a. The content of indole compounds in fruiting bodies of mushrooms from the *Basidiomycota* division [mg/100g d.w.]

Species / Indole compounds	*Agaricus bisporus*	*Boletus edulis*	*Cantharellus cibarius*	*Lactarius deliciosus*	*Lactarius deterrimus*
L-Tryptophan	0.39	0.39	0.01	<0.001	0.23
5-Hydroxy-L-tryptophan	<0.001	0.18	0.02	0.25	–
Tryptamine	0.06	1.17	0.01	<0.001	2.73
Melatonin	0.11	0.68	0.14	1.29	0.19
Serotonin	5.21	10.14	29.61	18.42	–
Indole	–*	–	–	–	0.62
Indole acetonitrile	<0.001	0.04	0.01	0.62	–
Indoleacetic acid	0.19	<0.001	<0.001	2.04	0.36
Kynurenic acid	6.21	<0.001	<0.001	<0.001	–
Kynurenic sulfate	<0.001	<0.001	4.81	39.2	–

(*): trace amounts found.

On the other hand, in the case of *Lactarius deterrimus* fruiting bodies, the newly developed TLC method with densitometric detection enabled determination of 5-methylcytosine, tryptamine, melatonin, tryptophan, indoleacetic acid and indole. The compound with the highest determined amount was tryptamine 2.72 mg/100 g d.w., and the total indole compound content in fruiting bodies was 4.88 mg/100 g d.w.

Table 2 b. The content of indole compounds in fruiting bodies of mushrooms from the *Basidiomycota* division [mg/100g d.w.]

Species / Indole compounds	*Leccinum rufum*	*Pleurotus ostreatus*	*Suillus luteus*	*Tricholoma equestre*	*Imleria badia*
L-Tryptophan	<0.001	<0.001	2.61	2.85	0.68
5-Hydroxy-L-tryptophan	0.02	2.08	1.63	0.59	<0.001
Tryptamine	1.05	0.91	<0.001	2.01	0.47
Melatonin	0.08	<0.001	0.71	<0.001	<0.001
Serotonin	31.71	6.52	34.11	0.18	0.52
Indole	–	–	–	–	–
Indole acetonitrile	0.45	0.01	<0.001	–	–
Indoleacetic acid	0.89	0.21	–	0.04	–
Kynurenic acid	<0.001	<0.001	2.63	<0.001	1.6
Kynurenic sulfate	<0.001	<0.001	19.57	<0.001	1.96

The difficulty in obtaining research material, associated with the temporary and frequently unpredictable emergence of wild growing mushroom fruiting bodies has been the reason to undertake actions to obtain theirs *in vitro* mycelial cultures. Under conditions of mycelial cultures, it is also less difficult to control the accumulation of the examined bioactive substances.

In methanol extracts from mycelial cultures biomass of the edible *Tricholoma equestre* (L.) P. Kumm and *Imleria badia* (Fr.) Fr. the content of indole derivatives has been determined. The conducted study has proven the applicability of mycelial culture as the source of the examined compounds.

The HPLC analysis with UV detection has demonstrated the presence of 8 compounds (L-tryptophan, 5-hydroxy-L-tryptophan, serotonin, melatonin, tryptamine, indole, indole acetamide and kynurenine sulfate

(VI)). Three common metabolites in fruiting bodies of both species were found: L-tryptophan, tryptamine, and serotonin; in the biomass from *in vitro* cultures 2 compounds: L-tryptophan and tryptamine. Contents of the determined compounds were highly variable, both extracts from fruiting bodies (0.18–2.85 mg/100 g d.w.), as well as from *in vitro* culture biomass (0.32–2.10 mg/100 g d.w.) (Table 2b.). The relatively high content of L-tryptophan, serotonin and tryptamine was noticeable [52].

Analysis of indole compound content in the fruiting bodies and mycelial cultures of *Cantharellus cibarius* demonstrated the considerable content of serotonin (17.61 mg/100 g d.w.; 20.49 mg/100 g d.w., respectively) and 5-hydroxy-L-tryptophan (similar amounts to 12.52 mg/100 g d.w.).

Methanol extracts from biomass of *in vitro* cultures and fruiting bodies of *Sarcodon imbricatus* L. P. Karst. were found to contain: L-tryptophan, tryptamine, and serotonin. In addition, extract from fruiting bodies was found to contain minor amounts of melatonin. The total content of indole compounds was 89.38 mg/100 g d.w. in fruiting bodies and 45 mg/100 g d.w. in mycelial cultures. Serotonin was the dominant compound in both tested extracts (52.02 mg/100 g d.w. and 3.03 mg/100 g d.w.– respectively). The identity of serotonin was confirmed with results of the conducted ^{1}HNMR spectral analysis [53].

In two other species of aphyllophoroidal mushrooms (sensu lato): *Hydnum repandum* L. and *Sparassis crispa* (Wulf.) Fr., the presence of 5 compounds was determined, that is indole, melatonin, serotonin, tryptamine, and L-tryptophan [54].

Another study attempted to determine, which of the *Agaricus bisporus* cultivars (white, brown, portobello) is characterized by higher nutritional and health-promoting value, and in addition to this objective, the purposefulness of removing the upper layer from mushroom caps prior to use as food. The studies utilized lyophilized parts of fruiting bodies (cap skin, cap, stems) of the white, brown and portobello cultivars of commercial origin.

The study demonstrated that *A. bisporus* fruiting bodies of the brown cultivar are most rich in indole compounds and it has been demonstrated that their highest amounts are accumulated in the mycelium located on the outer side of the fruiting body, that is the 'cap skin'. Differences in physiologically active contents of active organic compounds in individual parts of three *A. bisporus* cultivars have been shown. Due to their dietary value, removing the mushroom cap skin is unjustified [55].

Analysis of the indole compounds content in *Phellinus,* a genus of mushrooms growing on wood, demonstrated the presence of three metabolites (serotonin, tryptamine, and L-tryptophan). The content of individual indole compounds was from 1.70 (tryptamine in *P. robustus*) to 8.32 mg/100 g d.w. (L-tryptophan in *P. robustus*) [56].

The presence of indole compounds in methanol extracts was also analyzed in the biomass obtained from mycelial cultures and fruiting bodies of *Piptoporus betulinus.*

The biomass of *in vitro* cultures of *P. betulinus* was found to contain 5-hydroxy-L-tryptophan, 5-methyltryptamine, and L-tryptophan. The highest content characterized 5-methyltryptamine (3.9 μg/100 d.w.), while the lowest L-tryptophan (1.34 μg/100 d.w.). The total content of indole compounds in the biomass was 8.07 μg/100 d.w. Comparative analysis of the content of these compounds in fruiting bodies demonstrated total content at a similar level of 10.33 μg/100 d.w. (under review).

PRESENCE OF INDOLE COMPOUNDS IN BASIDIOMYCOTA FRUITING BODIES AFTER HEAT TREATMENT

Another research trend concerning indole compounds was to determine their content in fruiting bodies subject to heat treatment, simulating the procedure of cooking meals based on mushrooms, as they are rarely consumed raw, and indole compounds decompose under the influence of temperature (L-tryptophan is one of the most stable compounds in this group). L-tryptophan has been determined in the fruiting bodies of eleven species: *A. mellea, A. polytricha, I. badia, B. edulis, C. cibarius, L. deliciosus, L. scabrum, L. edodes, M. procera, P. ostreatus, and S. bovinus.* The highest diversity of indole compounds was determined in extracts from unprocessed *C. cibarius* fruiting bodies; methanol extract from this species was found to contain 7 out of 13 examined indole compounds. The species, in which six indole compounds were found are *B. badius, M. procera,* and P. *ostreatus*. The compounds occurring at the highest amounts in the tested mushrooms were: serotonin (29.61 mg/100 g d.w. in *C. cibarius*), L-tryptophan (to 25.90 mg/100 g d.w. in *S. bovinus*) and 5-hydroxy-L-tryptophan (24.83 mg/100 g d.w. in *L. edodes*). Fruiting bodies of *A. mellea* after heat treatment was found to contain the lowest amounts of indole compounds (only L-tryptophan and tryptamine) [5, 17].

Every tested species contained L-tryptophan. The content of this metabolite ranged from 0.01 to 25.90 mg/100 g d.w. L-tryptophan as essential amino acid must be provided to a human organism with food. It is a substrate for synthesis of such important compounds as neurotransmitters (e.g., dopamine, melatonin, serotonin, and adrenaline) or vitamins (niacin). The highest L-tryptophan content was determined in extracts from unprocessed *S. bovinus* fruiting bodies–25.90 mg/100 g d.w. The highest amount of L-tryptophan was found also in *C. cibarius* fruiting bodies. Among extracts of fruiting bodies subject to heat treatment, also the extract from the Jersey cow mushroom fruiting bodies contained the high amount of L-tryptophan–17.71 mg/100 g d.w. Therefore, the processed edible mushrooms, and in particular *S. bovinus*, constitute a good source of L-tryptophan in the human diet.

5-Hydroxy-L-tryptophan, the direct precursor of serotonin and melatonin was present in both unprocessed and heat-processed fruiting bodies. However, its highest content was determined in material not subject to heat treatment. Serotonin was determined at the highest level among the analyzed indole compounds in unprocessed *C. cibarius* fruiting bodies (29.61 mg/100 g d.w.), but it was not found in the heat-processed material.

The fruiting bodies subject to heat treatment were found to contain approximately 2-fold lower total content of indole compounds. The highest total content of indole compounds was determined in the fruiting bodies of *C. cibarius*, *L. deliciosus, S. bovinus*, both unprocessed as well as heat treated [5, 17].

The Release of Indole Compounds to Artificial Digestive Juices

By looking at the presence of indole compounds in numerous mushroom species, it seemed important from the therapeutic viewpoint, to estimate the actual amounts of these compounds available to the human organism. To this end, a Gastroel-2014 device was used (patent application no. P 417238) that imitated natural peristalsis through a special type of shaking, as well as enabling extraction under human body temperature conditions, i.e., 36.6°C to 37°C. The applied method enabled obtaining a maximum correlation between the *in vitro/in vivo* system, ensuring the best possible simulation of the digestion process, which then translated into the possibility to estimate the bioavailability of indole compounds from the mushroom material. The use of artificial digestive juices with precisely developed compositions additionally enabled the conduct of extraction that was as close as possible to the digestion process occurring in the human organism [57-59].

Fruiting bodies of popular species have been used in the research: *Agaricus bisporus* of commercial origin (supermarket) and collected from the natural habitat in mixed forests of southern Poland: *Imleria badia*, *Cantharellus cibarius*, as well as *Trichoma equestre*. The study material also consisted of lyophilized (under the same conditions as the fruiting bodies) mycelia from *in vitro* cultures of these species, conducted on liquid medium acc. to Oddoux [2, 19, 60, 61]. The study objective was to determine whether fruiting bodies of edible mushrooms, as well as their mycelial cultures, were conducted under laboratory conditions on correctly selected media may, in fact, constitute a nutritional source of indole compounds. Obtaining the best level of indole compound release from the biomass of *in vitro* cultures was particularly important, as it enabled development and standardization of the conditions for the culture accumulating these physiologically important compounds, which may have a future impact on the obtaining of enriched biomass, which may be used to develop a natural diet supplement.

The indole compounds identified during the conducted experiments included 5-hydroxy-L-tryptophan, serotonin, L-tryptophan, 5-methyltryptamine, 6-methyl-D,L-tryptophan, melatonin and tryptamine. It was confirmed that indole compounds release in an efficient manner both to artificial gastric juice as well as in the conditions of artificial intestinal juice [2, 19, 60, 61]. It should be emphasized that after extraction of *I. badia* fruiting bodies in artificial gastric juice independently of the incubation time, and in *in vitro* cultures of *A. bisporus* (particularly when enriched with zinc hydroaspartate), samples were found to contain considerable amounts of 5-methyltryptamine (from 33.41 to 35.54 mg/100 g d.w.), which is particularly significant in the case of persons taking MAO inhibitor drugs due to the possibility of interaction [19]. Moreover, it was demonstrated that enrichment of mycelial cultures with zinc salt in organic form, that is zinc hydroaspartate, is favorable in terms of the total content of indole compounds identified in *in vitro* cultures of *A. bisporus* and *I. badia* [2, 60].

Based on the conducted experiments it was demonstrated that edible mushroom fruiting bodies, and in particular their biomass from *in vitro* cultures release indole compounds to the solution of artificial gastric juice and artificial intestinal juice. Thus, they have been shown to be their good source for human organism [2, 19, 60, 61]. The species, in which the highest total amounts of indole groups have been identified is *A. bisporus* [2]. Furthermore, considering the fact, that fruiting bodies and *in vitro* cultures of edible mushroom species turned out to be good sources of L-tryptophan and 5-hydroxy-L-tryptophan, they may constitute a good alternative for red meat, which is of particular importance in the vegetarian diet. Moreover, these compounds are directly transformed into serotonin and melatonin in the brain, thus the conducted research confirms the potential anti-depressive effect of *A. bisporus*, *C. cibarius*, *I. badia,* and *T. equestre*, as well as their i*n vitro* cultures [2, 19, 60, 61].

CONCLUSION

Numerous compounds containing indole skeleton play a significant role in medicine. They are of significance for the normal functioning of the human organism, and in particular of the brain and the entire nervous system. They play a considerable role as substrates for the synthesis of new chemical compounds, included in numerous therapeutic preparations. There is a considerable amount of evidence confirming the influence of indole derivatives, on both animals as well as plant organisms. These compounds have a strong effect on the immune and nervous systems of mammals.

For the first time, the conducted analysis of indole compounds in edible mushroom species and mycelia from *in vitro* cultures demonstrates that the obtained results are of considerable cognitive and scientific value. Moreover, the obtained study results provide the possibility for future use of the developed biotechnological methods in practice to control the content of indole metabolites in mycelia from *in vitro* cultures.

It has been proven that therapeutic mushrooms are a material, which after heat treatment and consumption remain their good source. The use of artificial gastrointestinal tract model has additionally shown the probable behavior of mushroom material during the digestion process, which has facilitated the determination of the probable amounts of indole compounds that are bioavailable to the human organism.

REFERENCES

[1] Moro, C., Palacios, I., Lozano, M., D'Arrigo, M., Guillamón, E., Villares, A., Martínez, J. A. and García-Lafuente, A. (2012). Anti-inflammatory activity of methanolic extracts from edible mushrooms in LPS activated RAW 264.7 macrophages. *Food Chemistry,* 130: 350-355.

[2] Muszyńska, B., Kała, K., Sułkowska-Ziaja, K., Krakowska, A. and Opoka, W. (2016). *Agaricus bisporus* and its *in vitro* culture as a source of indole compounds released into artificial digestive juices. *Food Chemistry*, 199: 509-515.

[3] Roupas, P., Keogh, J., Noakes, M., Margetts, C. and Taylor, P. (2012). The role of edible mushrooms in health: Evaluation of the evidence. *Journal of Functional Foods*, 4: 687-709.

[4] Lindequist, U., Niedermeyer, T. H. J. and Jülich, W. D. (2005). The pharmacological potential of mushrooms. *Evidence-Based Complementary and Alternative Medicine*, 2: 285-299.

[5] Muszyńska, B., Sułkowska-Ziaja, K. and Ekiert, H. (2011). Indole compounds in fruiting bodies of some edible Basidiomycota species. *Food Chemistry*, 125: 1306-1308.

[6] Chen, S. Y., Ho, K. J., Hsieh, Y. J., Wang, L. T. and Mau, J. L. (2012). Contents of lovastatin, g-aminobutyric acid and ergothioneine in mushroom fruiting bodies and mycelia. *LWT-Food Science and Technology*, 47: 274-278.

[7] Cheung, P. C. K. (2013). Mini-review on edible mushrooms as source of dietary fiber: Preparation and health benefits. *Food Science and Human Wellness*, 2: 162-166.

[8] Chihara, G., Hamuro, J., Maeda, Y., Arai, Y. and Fukuoka, F. (1970). Fractionation and purification of the polysaccharides with marked antitumor activity, especially lentinan, from *Lentinus edodes* (Berk.) Sing. (an edible mushroom). *Cancer Research*, 30: 2776-2781.

[9] Muszyńska, B., Sułkowska-Ziaja, K., Łojewski, M., Opoka, W., Zając, M. and Rojowski, J. (2013). Edible mushrooms in prophylaxis and treatment of human diseases. *Medicina Internacia Revuo*, 101: 170-183.

[10] Patel, S. and Goyal, A. (2012). Recent developments in mushrooms as anti-cancer therapeutics: a review. *3 Biotech*, 2: 1-15.

[11] Zong, A., Cao, H. and Wang, F. (2012). Anticancer polysaccharides from natural resources: A review of recent research. *Carbohydrate Polymers*, 90: 1395-1410.

[12] Fujii, T., Maeda, H., Suzuki, F. and Ishida, N. (1978). Isolation and characterization of a new antitumor polysaccharide, KS-2, extracted from culture mycelia of *Lentinus edodes*. *The Journal of Antibiotics*, 31: 1079-1090.

[13] Huang, S. J., Lin, C. P. and Tsai, S. Y. (2015). Vitamin D_2 content and antioxidant properties of fruit body and mycelia of edible mushrooms by UV-B irradiation. *Journal of Food Composition and Analysis*, 42: 38-45.

[14] Teichmann, A., Dutta, P. C., Staffas, A. and Jägerstad, M. (2007). Sterol and vitamin D2 concentrations in cultivated and wild grown mushrooms: Effects of UV irradiation. *LWT-Food Science and Technolgy*, 40: 815-822.

[15] Cacchi, S. and Fabrizi, G. (2005). Synthesis and functionalization of indoles through palladium-catalyzed reactions. *Chemical Reviews*, 105: 2873-2920.

[16] Leung, A. Y. and Paul, A. G. (1968). Baeocystin and norbaeocystis: new analogs of psilocybin from Psilocybe baeocystis. *Journal of Pharmaceutical Sciences*, 57: 1667-1671.

[17] Muszyńska, B., Sułkowska-Ziaja, K. and Wójcik, A. (2013). Levels of physiologically active indole deratives in the fruiting bodies of some edible mushrooms (Basidiomycota) before and after thermal processing. *Mycoscience*, 54: 321-326.

[18] Wurst, M., Kysilka, R. and Flieger, M. (2002). Psychoactive tryptamines from basidiomycetes. *Folia Microbiologica*, 47: 3-27.

[19] Muszyńska, B., Kała, K., Sułkowska-Ziaja, K., Gaweł, K., Zając, M. and Opoka, W. (2015). Determination of indole compounds released from selected edible mushrooms and their biomass to artificial stomach juice. *LWT – Food Science and Technology*, 62: 27-31.

[20] Muszyńska, B., Komendacki, P., Kała, K., Opoka, W. and Rojowski, J. (2014). L-Tryptophan and its derivatives in edible mushrooms species. M*edicina Internacia Revuo*, 2(103), 82-88.

[21] Kohlmünzer, S. (2012). *Farmakognozja. Podręcznik dla studentów farmacji*. [*Pharmacognosy. Handbook for pharmacy students*]. PZWL, Poland: Warszawa.

[22] Liu, J. (2005). N-Containing of Macromycetes. *Chemical Reviews*, 105: 2723-274.

[23] Wurst, M., Kysilka, R. and Koza, T. (1992). Analysis and isolation of indole alkaloids of fungi by high-performance liquid chromatography. *Journal of Chromatography*, 593, 201-208.

[24] Ott, J. and Guzman, G. (1976). Detection of psilocybin in species of Psilocybe, Panaeolus and Psathyrella. *Lloydia*, 39: 258-260.

[25] Buczko, P., Cylwik, D. and Stokowska, W. (2005). Metabolizm tryptofanu w ślinie szlakiem kynureinowym. *Postępy Higieny i Medycyny Doświadczalnej*, 59: 283-289. [Metabolism of tryptophan via the kynurenine pathway in saliva. *Advances in Hygiene and Experimental Medicine*, 59: 283-289].

[26] Kaneez, F. S. and Arshad, S. S. (2007). The metabolism of serotonin in neuronal cells in culture and platelets. *Experimental Brain Research*, 183: 411-416.

[27] Stone, T. W., Mackay, G. M., Forrest, C. M., Clark, C. J. and Darlington, L. G. (2003). Tryptophan metabolites and brain disorders. *Clinical Chemistry and Laboratory Medicine*, 41: 852-859.

[28] Stępień, A., Walecka-Kapica, E., Błońska, A. and Klupińska, G. (2014). Rola tryptofanu i serotoniny w patogenezie i leczeniu zespołu jelita nadwrażliwego. Folia Medica Lodziensia, 41: 139-154. [The role of tryptophan and serotonin in pathogenesis and treatment of irritable bowel syndrome. *Folia Medica Lodziensia*, 41: 139-154].

[29] Keszthelyi, D., Troost, F. J. and Masclee, A. M. (2009). Understanding the role of tryptophan and serotonin metabolism in gastrointestinal function. *Neurogastroenterology and Motility*, 21:1239-1249.

[30] Turner, E. H., Loftis, J. M. and Blackwell, A. D. (2006). Serotonin a la carte: Supplementation with the serotonin precursor 5-hydroxytryptophan. *Pharmacology & Therapeutics*, 109: 325-338.

[31] Birdsall, T. C. (1998). 5-Hydroxytryptophan: A Clinically-Effective Serotonin Precursor. *Alternative Medicine Review*, 3: 271-280.

[32] Baran, H., Jellinger, K. and Deecke L. (1999). Kynurenine metabolism in Alzheimer's diseases. *Journal of Neural Transmission*, 106: 165-181.

[33] Ouchi, Y., Yoshikawa, E., Futatsubachi, M., Yagi, S., Ueki, T. and Nakamura, K. (2009). Altered brain serotonin transporter and associated glucose metabolism in Azheimere disease. *Journal of Nuclear Medicine*, 50: 1260-1266.

[34] Esposito, E. and Cuzzocrea, S. (2010). Antiinflammatory Activity of Melatonin in Central Nervous System. *Current Neuropharmacology*, 8: 228-242.

[35] Singhal, N. K., Srivastava, G., Agrawal, S., Jain, S. K. and Singh, M. P. (2012). Melatonin as a neuroprotective agent in the rodent models of Parkinson's disease: is it all set to irrefutable clinical translation? *Molecular Neurobiology*, 45: 186-199.
[36] Morgan, Y. R., Turner, P., Kennedy, B. J., Hambley, T. W., Lay, P. A., Biffin, J. R., Regtop, H. L. and Warwick, B. (2001). Preparation and characterization of dinuclear copper-indomethacin anti-inflammatory drugs. *Inorganica Chimica Acta*, 324: 150-161.
[37] Paredes, S. D., Korkmaz, A., Manchester, L. C., Tan, D. X. and Reiter, R. J. (2009). Phytomelatonin: a review. *Journal of Experimental Botany*, 60: 57-69.
[38] Reiter, R. J., Tan, D. X., Burkhardt, S. and Manchester L. C. (2001). Melatonin in Plants. *Nutrition Reviews*, 59: 286-290.
[39] Wang, X. (2009). The anti-apoptotic activity of melatonin in neurodegenerative diseases. *CNS Neuroscience & Therapeutics*, 15: 345-357.
[40] Jones, R. S. (1982). Tryptamine: a neuromodulator or neurotransmitter in mammalian brain? *Progress in Neurobiology*, 19: 117-139.
[41] Juorio, A. V. and Durden, D. A. (1984). The distribution and turnover of tryptamine in the brain and spinal cord. *Neurochemical Research*, 9: 1283-1293.
[42] Martin, W. R. and Sloan, J. W. (1970). Effects of infused tryptamine in man. *Psychopharmacology*, 18: 231-237.
[43] Martin, W. R., Sloan, J. W., Christian, S. T. and Clements, T. H. (1972). Brain levels of tryptamine. *Psychopharmacology*, 24: 331-346.
[44] Dobbs, A. (2001). Total synthesis of indoles from *Tricholoma* species via Bartoli/heteroaryl radical methodologies. *The Journal of Organic Chemistry*, 66: 638-641.
[45] Vanhaelen-Fastre, R. and Vanhaelen, M. (1984). Qualitative and quantitative determinations of hallucinogenic components of psilocybe mushrooms by reversed-phase high-performance liquid chromatography. *Journal of Chromatography*, 312: 467-472.

[46] Gartz, J. (1994). Extraction and analysis of indole derivatives from fungal biomass. *Journal of Basic Microbiology*, 34: 17-22.

[47] Grzybek, J. and Kohlmünzer, S. (1972). Charakterystyczne składniki chemiczne grzybów wielkoowocnikowych (Macromycetes) - cz.1 Substancje azotowe. [Characteristic chemical components of high friction fungi (Macromycetes) - part 1. Nitrogen substances.] *Wiadomości Botaniczne*, 16: 35-55.

[48] Sewerynek, E., Stuss, M., Oszczygieł, K., Kułak, J. and Lewiński, A. (2005). Protective effect of indole compounds on lipopolysaccharide induced lipid peroxidation in *in vitro* conditions. *Polish Journal of Endocrinology*, 4: 508-511.

[49] Turek, M. and Łodyga-Chruścińska, E. (2008). Zastosowanie w medycynie pochodnych indolu i jego kompleksów z biometalami. *Chemia Spożywcza i Biotechnologia*, 72: 73-88. [Application in medicine of indol the derivatives and its complexes with biometals. *Biotechnology and Food Science*, 72: 73-88].

[50] Muszyńska, B., Sułkowska-Ziaja, K. and Ekiert, H. (2011). Indole Compounds in Some Culinary-Medicinal Higher Basidiomycetes from Poland. *International Journal of Medicinal Mushrooms*, 13: 449-454.

[51] Muszyńska, B., Sułkowska-Ziaja, K. and Ekiert, H. (2011). Analysis of indole copounds in fruiting bodies of *Armillaria mellea*. *Acta Poloniae Pharmaceutica – Drug Research*, 68: 93-97.

[52] Muszyńska, B., Sułkowska-Ziaja, K. and Ekiert, H. (2009). Indole compounds in fruiting bodies of some selected Macromycetes species and in their mycelia cultured *in vitro*. *Die Pharmazie*, 64: 479-480.

[53] Sułkowska-Ziaja, K., Muszyńska, B. and Ekiert, H. (2013). Analysis of indole com- pounds from the fruiting bodies and the culture mycelia of *Sarcodon imbricatus*. *Mycoscience*, 55: 164-167.

[54] Sułkowska-Ziaja, K., Muszyńska, B. and Szewczyk, A. (2015). Antioxidant components of selected indigenous edible mushrooms of the obsolete order Aphyllophorales. *Revista Iberoamericana de Micología*, 32: 99-102.

[55] Muszyńska, B., Piotrowska, J., Krakowska, A., Gruba, A., Kała, K., Sułkowska-Ziaja, K., Kryczyk, A. and Opoka, W. (2017). Study of physiologically active components in different parts of fruiting bodies in varieties of *Agaricus bisporus* (white mushroom). *European Food Research and Technology*, 243: 2135-2145.

[56] Sułkowska Ziaja, K., Maślanka, A., Szewczyk, A. and Muszyńska, B. (2017). Physiologically active compounds in four species of Phellinus. *Natural Product Communications*, 12: 363-366.

[57] Arvidson, K. and Johasson, E. G. (1985). Galvanic current between dental alloys *in vitro*. *Scandinavian Journal of Dental Research*, 93: 467-473.

[58] Neumann, M., Goderska, K., Grajek, K. and Grajek, W. (2006). Modele przewodu pokarmowego *in vitro* do badań nad biodostępnością składników odżywczych. *ŻYWNOŚĆ. Nauka. Technologia. Jakość*, 1: 30-45. [The *in vitro* models of gastrointestinal tract to study bioavailability of nutriments. *Food. Science. Technology. Quality*, 1: 30-45].

[59] *Polish Pharmakopeia (10th ed.)* (2014). PTFarm, Poland: Warszawa.

[60] Kała, K., Maślanka, A., Sułkowska-Ziaja, K., Rojowski, J., Opoka, W. and Muszyńska, B. (2016). *In vitro* culture of *Boletus badius* as a source of indole compounds and zinc ions released in artificial digestive juices. *Food Science and Biotechnology*, 25: 829-837.

[61] Kała, K., Sułkowska-Ziaja, K., Rojowski, J., Opoka, W. and Muszyńska, B. (2016). *Tricholoma equestre* species as a source of indole compounds and zinc released into artificial digestive juices. *Medicina Internacia Revuo*, 27: 35-39.

In: Indole: Synthesis, Functions and Reactions ISBN: 978-1-53614-777-3
Editor: Bridget Pratt

Chapter 3

SYNTHETIC STRATEGIES TO PYRIMIDO [4,5-*B*]INDOLE: A PRIVILEGED FUSED-INDOLE SCAFFOLD

Claudia Kim, Ming-Ming Zhou and Nilesh Zaware [*]
Department of Pharmacological Sciences,
Icahn School of Medicine at Mount Sinai, New York, NY, US

ABSTRACT

Pyrimido-indoles are tricyclic ring systems produced when an indole group is fused with a pyrimidine ring. Of the two possible isomers of this ring system, the pyrimido[4,5-*b*]indoles are of particular interest relative to pyrimido[5,4-*b*]indoles, due to their presence in a wide range of pharmacologically active molecules. Compounds containing the pyrimido[4,5-*b*]indole chemotype have demonstrated affinity towards multiple anticancer biological targets including receptor tyrosine kinases, thymidylate synthase, BET bromodomains, and tubulin. In addition, derivatives bearing this scaffold exhibit antiviral, diuretic, anti-opportunistic infection, and hematopoietic stem-cell expansion activities. Multiple synthetic approaches to construct this notable tricycle have been

[*] Corresponding Author Email: nilesh.zaware@mssm.edu.

reported including the Nenitzescu reaction, Fisher-type cyclization, palladium catalysis, and aza-Wittig type reaction among others. This book chapter provides an overview of the diverse synthetic approaches to generate pyrimido[4,5-*b*]indoles.

Keywords: pyrimido[4,5-*b*]indole, multicomponent reaction, tricycle, organic synthesis

1. Introduction

Pyrimido-indoles are fused ring systems that can exist in two isomeric forms: Pyrimido[4,5-*b*]indole **1** and Pyrimido[5,4-*b*]indole **2** (Figure 1). The Pyrimido[4,5-*b*]indole chemotype has received tremendous attention from medicinal chemists due to its presence in a broad range of pharmacologically active compounds. Molecules featuring cores structure **1** have been shown to possess BET bromodomain inhibition [1], antiviral [2], stem-cell expansion [3], antiangiogenic [4], microtubule disruption [5], antiopportunistic infection [6] activities among others. This book chapter provides an overview of established protocols to construct this previliged scaffold.

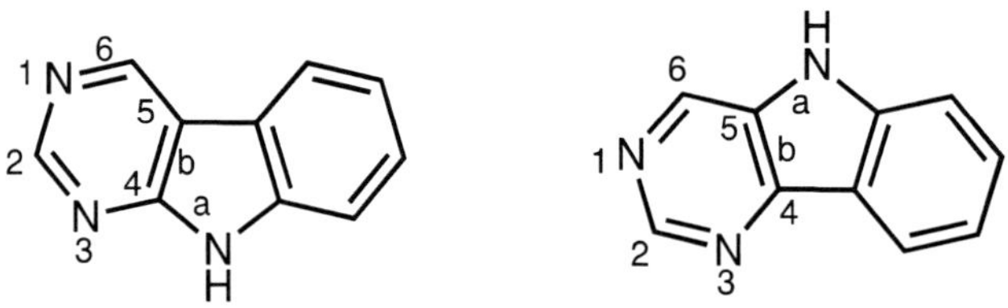

1 Pyrimido[4,5-*b*]indole

2 Pyrimido[5,4-*b*]indole

Figure 1. Two isomeric forms of pyrimido[4,5-*b*]indole.

2. Synthetic Strategies to Pyrimido[4,5-*b*]indoles

Synthesis to substituted pyrimido[4,5-*b*]indoles can be generally classified in the three categories below based on the starting materials used.

2.1. From Substituted Pyrimidines

Zhang et al. [7] formulated an optimized reaction condition to synthesize various pyrimido[4,5-*b*]indoles from pyrimidine substrates using palladium catalyzed arylation (Scheme 1). Several 4-anilino-5-iodopyrimidines **4** (R_1Ph = 4-EtPh, 4-CO_2MePh, 4,5- diCO_2MePh, 3,4,5-triOMePh, 4-SO_2NMe_2Ph, 3-BrPh, 4-Br, 2-MePh, 3-(2-oxazolyl)Ph; 5-Indolyl; R_2N-PhR_1 = 4-methoxy-*N*-methylaniline) in good yields (75-89%) were prepared from 4-chloro-5-iodopyrimidine **3** and substituted anilines in refluxing EtOH. Compound **4** was subjected to 10% mol $Pd(OAc)_2(PPh_3)_2$ and NaOAc (1.5 eq.) in 0.01M DMF at 85 °C to form the cyclic products (**5**) in yields ranging from 26 to 74%.

Dotzauer et al. [8] utilized an extension of Nenitzescu reaction to prepare pyrimido[4,5-*b*]indoles via a one pot procedure from pyrimidine-2,4,6-triamine (Scheme 2). At method development stage, pyrimidine-2,4,6-triamine **7a** and 1,4-benzoquinone **6** were combined in boiling glacial acetic acid to give a dark solution. Compound **11a** was isolated as a single product from the solution in 32% yield. As per the proposed mechanism, addition of pyrimidine-2,4,6-triamine **7a** to the enone substructure of **6** resulted in a Michael adduct (**8a**) formation. Compound **8a** was then oxidized by **6** to a quinone intermediate **10a**, which underwent an intramolecular cyclization and a final hydrogenation by hydroquinone **9** which formed 2,4-diaminopyrimido[4,5-*b*]indol-5-ol (**11a**). The 6-aminopyrimidines (**7b-f**) were treated with **6** under the same conditions to furnish the pyrimido[4,5-*b*]indoles **11b-f**. The title compounds differed in substituents at the 2 and 4 positions. It was found that 2,6-diamino-4-hydroxypyrimidine (**7h**) and its methylated congeners **7i** and **7j** also reacted with **6** affording compounds **11h-j** in moderate yields.

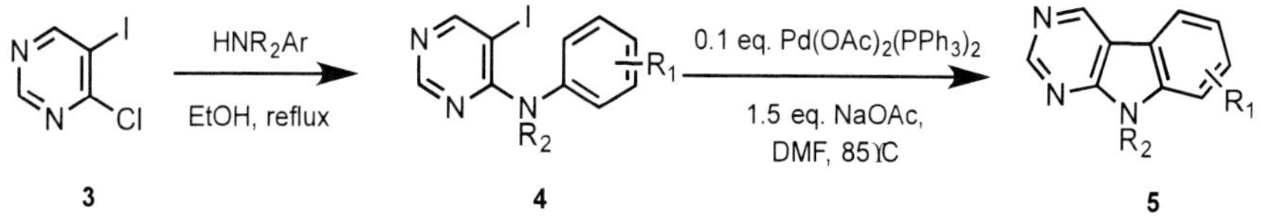

Scheme 1. Synthesis of pyrimido[4,5-*b*]indoles via a palladium –catalyzed arylation of pyrimidine substrates.

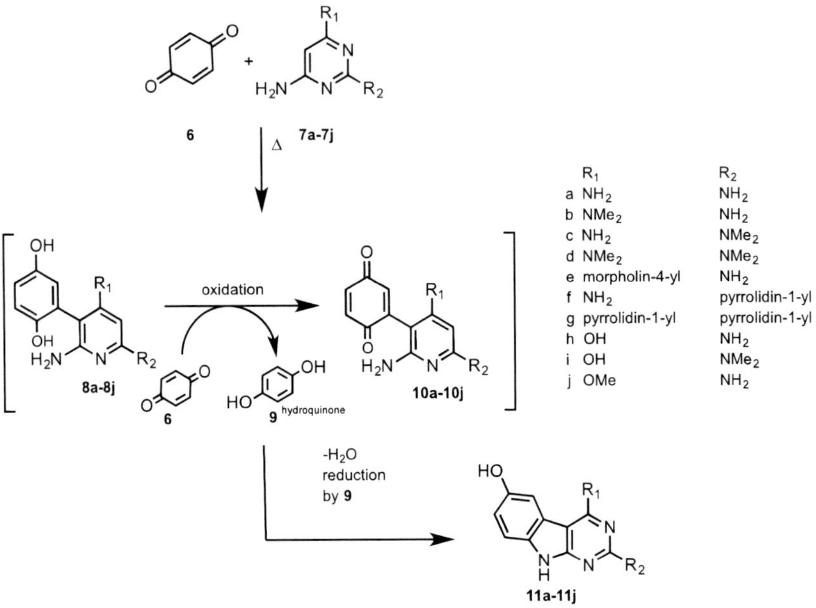

Scheme 2. Preparation of pyrimido[4,5-*b*]indoles using the Nenitzescu reaction.

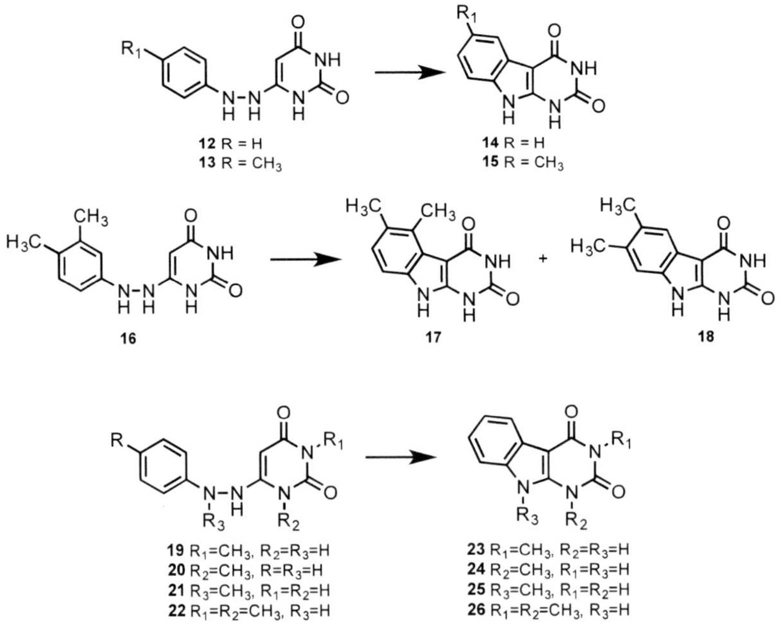

Scheme 3. Preparation of pyrimido[4,5-*b*]indoles via a facile Fischer-type cyclization.

G. E. Wright et al. [9] have been able to synthesize pyrimido[4,5-*b*]indoles via Fischer-type cyclization of 6-(phenylhydrazino)uracils in either formic acid or *N* hydrochloric acid at reflux (Scheme 3). The expected pyrimidoindoles were obtained after reflux of *C*-methyl-6-(phenylhydrazino) uracils in either formic acid or *N* hydrochloric acid for 1 hour. With the exception of **14** (from **12**), the formation of pyrimido[4,5-*b*]indole was always high yielding in hydrochloric acid than in formic acid. However, the reactions in formic acid were always complete as indicated by consumption of starting material. It was learned that *C*-methyl substitution played a role in reaction yields: the yield of **14** in hydrochloric acid was 21%, compared to the yield of **15** (R = CH_3) which was 64%, and the dimethyl isomer mixture (**17**, **18**) was 84%. The reaction parallels the classical Fischer indole synthesis. The probable mechanism begins with an *N*-protonation, followed by nucleophilic attack by an *ortho* phenyl carbon to the uracil 5-carbon with cleavege of the *N-N* bond. A hydrogen ion is lost followed by tautomerization resulting in a diamine, which would undergo ring closure by expulsion of the uracil amino group. The uracil amino group is displaced in the final step and is clear from the formation of compounds **25** (R" = CH_3, R = R' = H) from **21** (R" = CH_3, R = R' = H), where the *N*-methyl group is retained in the product.

Lapachev et al. [10] reported the formation of pyrimido[4,5-*b*]indoles from 4-azidoouracils (Scheme 4). The synthesis was initiated by conversion of 1,3-disubstituted barbituric acids (**27a** and **27b**) to 4-chlorouracils (**28a** and **28b**) with phosophorylchloride. Reaction of chlorouracils in dimethylformamide with sodium azide afforded azidouracil compounds (**29a** and **29b**). Thermolysis of compounds **29a** and **29b** in boiling xylene afforded the fused pyrimido[4,5-*b*]indoles **30** in excellent yields.

Intermediate compounds with the methyl series could be obtained as crystalline substances but chloro compound **28b** and **29** from the benzyl series could only be isolated as oil. In addition, obtaining dibenzyl barbituric acid (**27b**) in crystalline form was proven difficult, which was the reason to synthesize the pyrimido-indoles (**30**) without isolating any intermediate compounds.

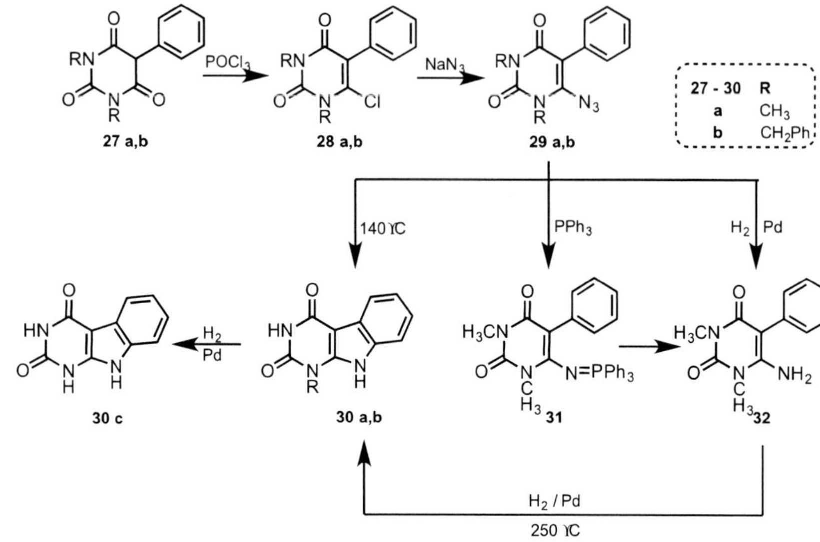

Scheme 4. Synthesis of pyrimido[4,5-*b*]indoles from 4-azidoouracils.

Synthesis was initiated with commercially available diethyl phenylmalonate, which was then isolated to the desired pyrimido-indoles with an overall yield of 50%. Unsubstituted pyrimido[4,5-*b*]indoledione (**30c**) was acheived with hydrogenolytic debenzylation of the dibenzyl indole **30b**.

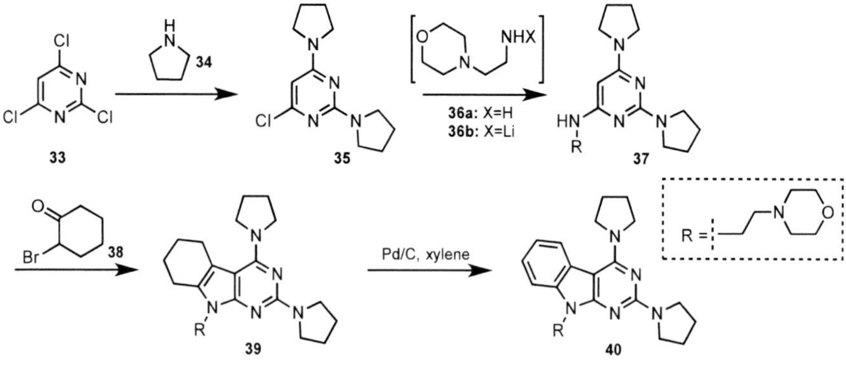

Scheme 5. Preparation of pyrimido[4,5-*b*]indoles from tri-substituted pyrimidines.

Iminophosphorane **31** is formed when reacting triphenylphospane with azide **29a**, the iminophosphorane can then be hydrolyzed to the aminouracil **32**. Another method to obtaining the amine **32** is catalytic reduction of the azide **29a**.

Aminouracil **32** can also undergo cyclohydrogenation in boiling diphenylether with the presence of Pd/H_2 catalyst to achieve the indole ring closure to pyrimido[4,5-*b*]indolediones **30a**.

Mauragis et al. [11] developed a large-scale synthesis that produced > 100 kg of compound **40** (Scheme 5) from tri-substituted pyrimidines. The synthesis was initiated with a selective heterogeneous double displacement of two chloro groups in trichloropyrimidine **33** by pyrrolidine to furnish **35**. High temperatures and/or long reaction times are required for the displacement of chlorine in **35** when heating the reaction with 4-(2-aminoethyl)-morpholine (**36a**). The large-scale reaction (1 Kg scale) created an exotherm that was difficult to control. This hurdle was bypassed by the activation of 4-(2-aminoethyl)morpholine to lithium amide (**36b**) that allowed the displacement of the 6-chloro in **35** to occur at reasonable temperatures to afford **37**. Intermediate **37** was converted to the cyclhexyl-fused derivative **39** via alklylation/cyclization with bromocyclohexanone **38** and finally the pyrimido[4,5-*b*]indole **40** was formed after dehydrogenation with 5%Pd/C in xylene.

2.2. From Substituted Pyrroles

Traxler et al. [12] reported the synthesis of 4-(phenylamino)-9*H*-pyrimido[4,5-*b*]indole pyrrolopyrimidine **46** in 1996 (Scheme 6). Starting compound 2-hydroxycyclohexanone **41** was reacted with benzylamine and malononitrile to 2-amino-1-benzyl-4,5,6,7-tetrahydro-1*H*-indole-3-carbonitrile **42**. The compound was then refluxed in 85% formic acid which afforded 9-benzyl-6,7,8,9-tetrahydro-5*H*-pyrimido[4,5-*b*]indole-4-ol **43** which was then coverted to corresponding chloride **44** using phosphoryl chloride. Replacement of the chlorine with corresponding substituted aniline afforded 4-(phenylamino)-5,6-substituted-7-benzylpyrrolo[2,3-*d*]pyrimidines. The resulting final product was obtained after the removal of the benzyl protecting group via oxidation of the N-protected compound **45** with DDQ.

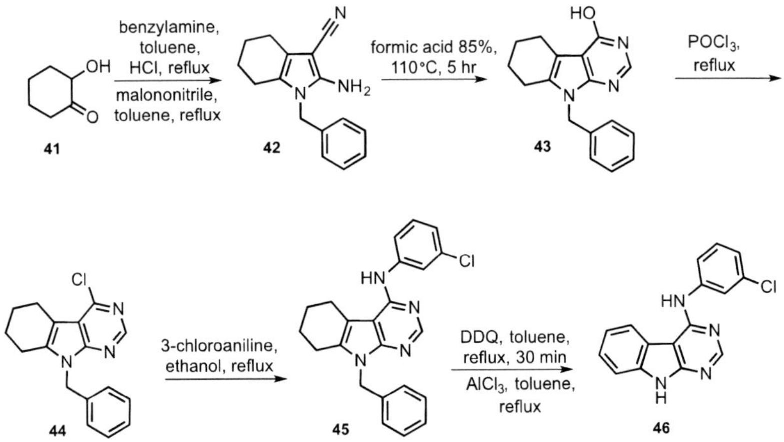

Scheme 6. Preparation of pyrimido[4,5-*b*]indoles from 2-hydroxycyclohexanone.

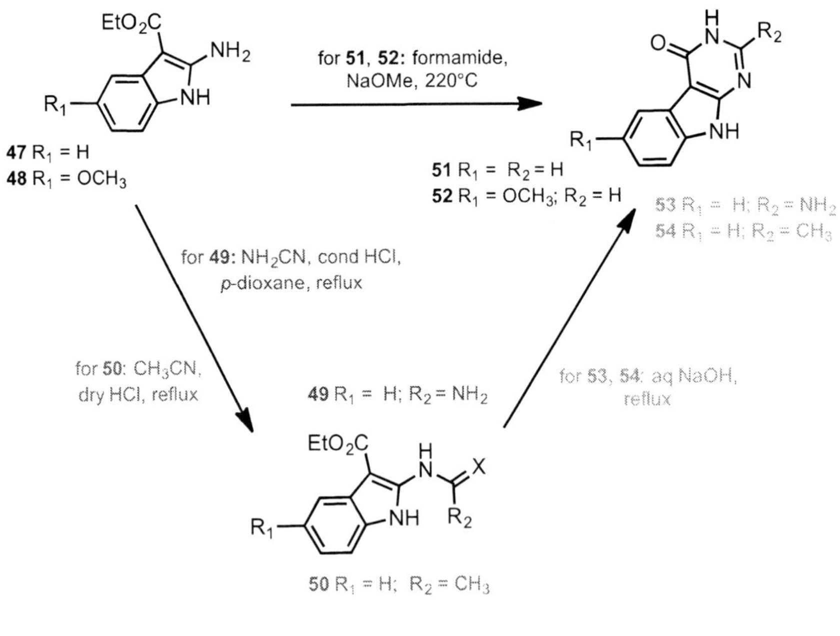

Scheme 7. Preparation of pyrimido[4,5-*b*]indoles from 2-amino-1*H*-indole-3-carboxylic acid ethyl ester.

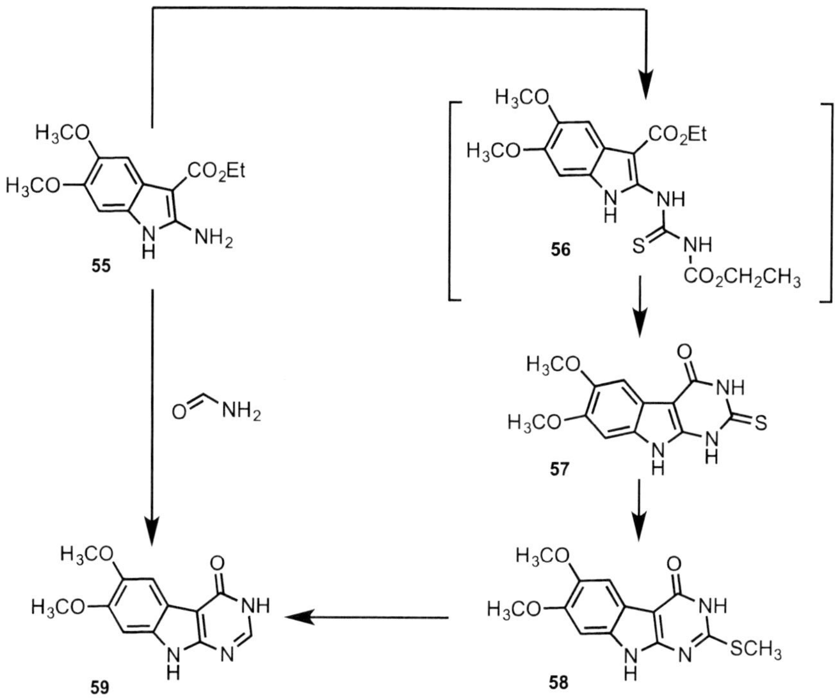

Scheme 8. Formation of pyrimido[4,5-*b*]indoles using the enamino ester ethyl 5,7-dimethoxy-2-amino-3-indolecarboxylate.

Showalter et al. [13] reported a versatile procedure to synthesize tricyclic compounds from either 2-amino-1*H*-indole-3-carboxylic acid ethyl ester **47** or its 6-methoxy congener **48**. The tricyclic compounds obtained by this method (**51-54**) are shown in Scheme 7. Compounds **51** and **52** were synthesized via ring closure of **47** and **48** respectively at 220°C with formamide and sodium methoxide. Condensation of **47** with cyanamide with concentrated HCl resulted in the guanidine **49**. Compound **49** was then refluxed with sodium hydroxide for ring closure reaction to 2-amino-1,9-dihydro-4*H*-pyrimido[4,5-*b*]indole-4-one **53**. Condensation of **47** with acetonitrile resulted in compound **50** followed by a similar ring closure reaction to form **54**.

Venugopalan et al. [14] synthesized 5,7-dimethoxypyrimido[4,5-*b*]indole from enamino ester ethyl 5,7-dimethoxy-2-amino-3-indolecarboxylate as a synthon (Scheme 8). Starting compound **55** was treated with ethyl carboxyisothiocyanate followed by a reaction with potassium hydroxide resulting in 2-thio-4-oxo-pyrimido[4,5-*b*]indole (**57**). Compound **58** was obtained via controlled methylation of **57**. Pyrimido indole **59** was was obtained by reduction of **58** with Raney Nickel. Alternatively, compound **59** was also directly obtained by condensation of the enamino ester **55** with formamide.

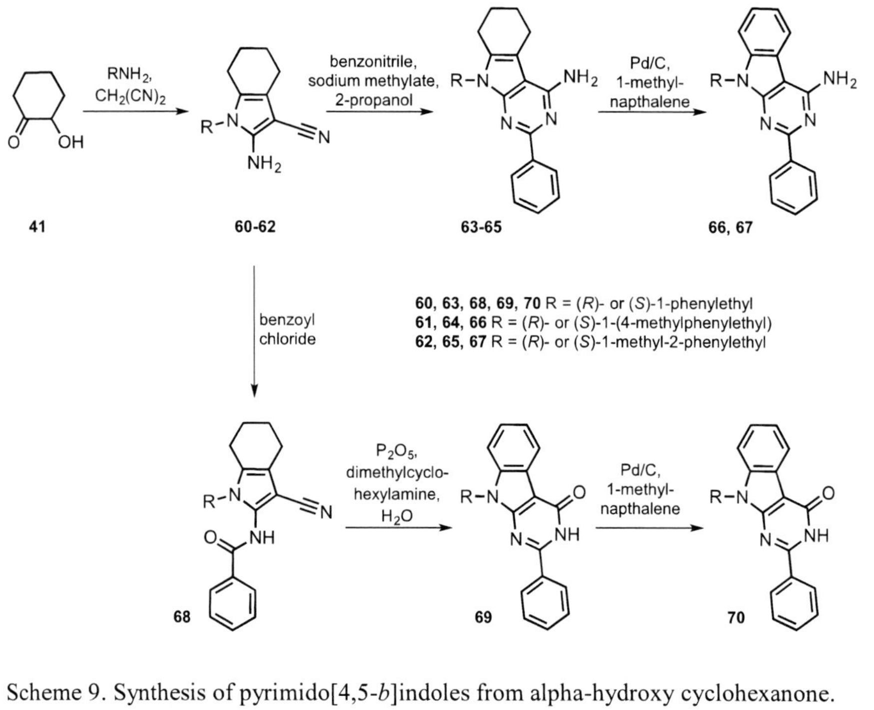

Scheme 9. Synthesis of pyrimido[4,5-*b*]indoles from alpha-hydroxy cyclohexanone.

Müller et al. [15] synthesized a series of chiral pyrimido[4,5-*b*]indole derivatives **60-70** (Scheme 9). The synthesis was initiated with the condensation of hydroxycyclohexanone **41** with chiral arylalkylamine in the presence of *p*-toluenesulfonic acid followed by base-catalyzed cyclization with malonitrile to obtain the desired products **60-62**. Enantiomers of compounds **60** (R = *(R)*- or *(S)*-1-phenylethyl), **61** (R = *(R)*- or *(S)*-1-(4-methylphenylethyl), and **62** (R = *(R)*- *(S)*-1-methyl-2-phenylethyl) were prepared starting from the *R*- or *S*- configured amines respectively. Compounds **60**, **61**, and **62** were reacted with benzonitrile and sodium methylate in 2-propanol to yield 2-phenyltetrahydropyrimido[4,5-*b*]indoleamine derivatives **63-65**. Condensation of compound **60** with benzoyl chloride afforded compounds **68**. Cyclization of compound **68** with phosphorous pentoxide in the presence of a mixture of dimethylcyclohexylamine and water at 190°C yielded tricyclics **69**. Treatment of similar compounds with polyphosphoric acid at 70°C had led to dealkylation, however the phenylethyl moeity at N9 was still stable in extreme conditions. Dehydrogenation of tetrahydroindoles **64**, **65**, and **69** to form the desired aromatic indole derivatives **66**, **67**, and **70** was done catalytically with Pd/C at high temperatures and 1-methylnaphtalene as the suitable solvent for the reaction.

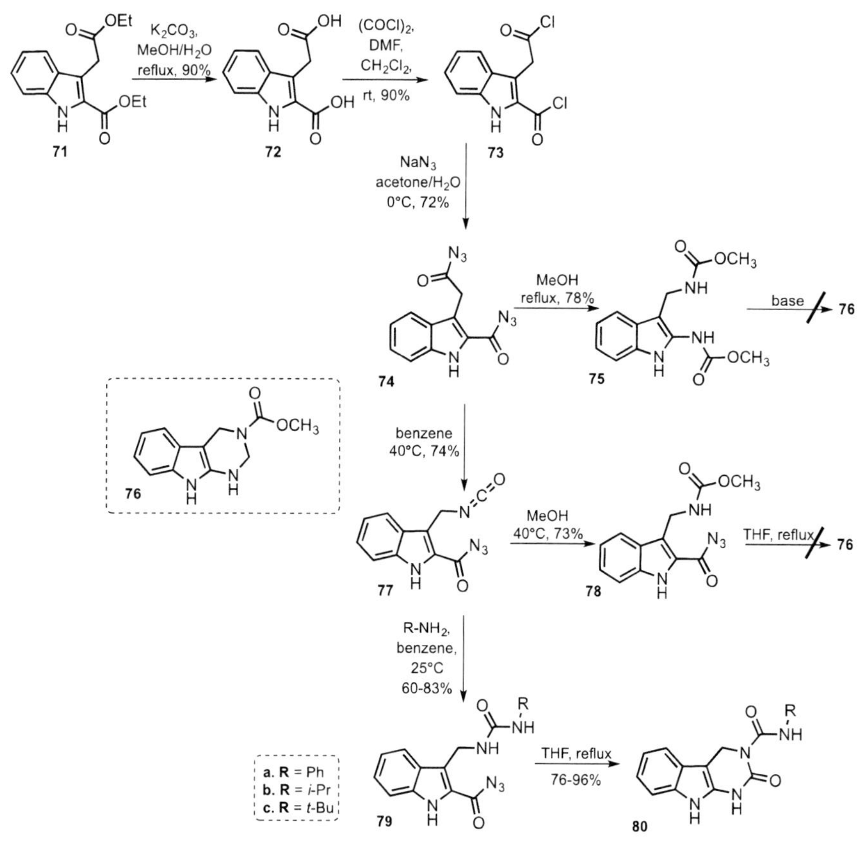

Scheme 10. Preparation of pyrimido[4,5-*b*]indoles via intramolecular cyclizations.

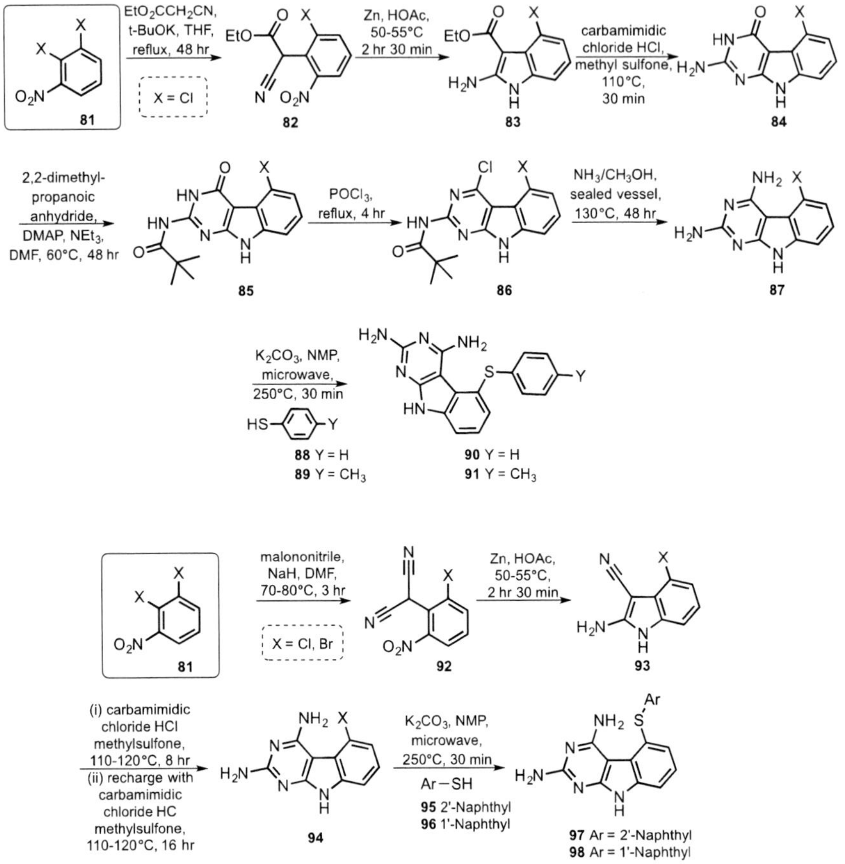

Scheme 11. Synthesis of pyrimido[4,5-*b*]indoles from 1,2-dichloro-3-nitrobenzene and 1,2-dibromo-3-nitro-benzene.

Kapti et al. [16] have reported a route for synthesis of pyrimido[4,5-*b*]indoles from an Indole 2,3-diester. There are two steps to the reported synthesis: (i) preparation of starting material (diester **71**) and (ii) transformation of **71** to tricyclic product (Scheme 10). The synthesis was initiated with the use of potassium carbonate in methanol/water to convert the diester **71** into diacid **72**. The diacid was then treated with oxalyl chloride in the presence of catalytic amount of *N,N*-dimethylformamide, followed by treatment with sodium azide in acetone/water to give diacyl azide **74**. Compound **74** was then converted to diurethane (**75**) in 78% yield via relux in methanol. The authors attempted to synthesize molecule **76** from the diurethane via intramolecular ring closure but failed. Compound **74** was heated in benzene at 35-40°C for 48 hours to yield **77** that had both acyl azide and isocyanate functionalities. Due to the strong electrophilicity of the monoisocyanate in **77**, both alcohol and amine-type nucleophiles were able to trap the structure to form urethane **78** and the urea derivatives **79a-c**, respectively. Thermal cyclization of **78** to **76** was unsuccessful. Urea derivatives (**79a-c**) underwent thermal cyclization that afforded the target compounds **80a-c** in high yields. The nucleophilicity of the NH groups in the urea derivatives is greater than that of urethane, which may have resulted in the successful thermal cyclizations.

In the past, Gangjee et al. [17] have been able to synthesize pyrimido[4,5-*b*]indoles in six steps from starting compound 2,3-dichloronitrobenzene (Scheme 11). The synthesis was initiated with the condensation of 1,2-dichloro-3-nitro-benzene **81** with potassium salt of ethyl cyanoacetate anion provided adduct 2-(2-chloro-6-nitrophenyl)-2-cyanoacetate **82**. Following the Zinc reduction of **82**, compound **83** (2-amino-4-chloro-1*H*-indole-3-carboxylate) was formed. Compound **83** was converted to 2-amino-5-chloro-3*H*-pyrimido[4,5-*b*]indol-4(9*H*)-one (**84**) via cyclo-condensation with carbamimidic chloride hydrochloride. The 2-amino group of compound **84** was protected using 2,2-dimethyl propanoic acid anhydride, which resulted in compound **85**. Treatment of compound **85** with phosphorous oxychloride at reflux afforded the 4-chloro derivative **86**. The 4-chloro group was then displaced with ammonia (sealed vessel reaction) and concomitant deprotection of the 2-amino group of compound **86** was achieved via a sealed-vessel reaction which gave 2,4-diamino compound **87**. Compound **87** was treated with substituted benzenethiols **88**, **89** in a microwave apparatus that resulted in the target compounds **90**, **91** respectively.

More recently, Zaware et al. [18] synthesized target pyrimido[4,5-*b*]indoles from the same starting compound 2,3-dichloronitrobenzene **81** mentioned before but in only four steps, which is 3 steps lower than the previously reported 7 step synthesis by Gangjee et al. [17]. This route can also be applied to the conversion of 1-2-dibromo-3-nitro-benzene to the target compound, extending the scope for the synthesis of pyrimido[4,5-

b]indoles. The recent protocol provides a less toxic, efficient, and more accomodating route to access pyrimido[4,5-*b*]indoles. The synthesis was initiated with the sodium hydride induced displacement 2,3-dichloronitrobenzene **81** with malononitrile to compound **92**, followed by Zn-dust catalyzed reduction to compound **93**. Initial cyclocondensation of compound **93** with carbamimidc chloride hydrochloride in methylsulfone (110-120°C for 8 hours) resulted in an incomplete reaction. The reaction mixture was recharged with carbamimic chloride hydrochloride with an additional 16 hours to the reaction time affording compound **94** in 45% yield (total reaction time of 24 hours). The net yield of this reaction route to common intermediate **94** was 15% over six steps, which is a 3-fold improvement over the previously reported synthesis also to compound **94** that had a net yield of 5% over six steps. The shorter route also avoided the use of toxic $POCl_3$ and the operationally difficult sealed-vessel reaction.

2.3. Miscellaneous Methods

Adib et al. [19] reported a new facile method for the synthesis of pyrimido[4,5-*b*]indoles from readily available building blocks such as oxindoles (Scheme 12). The initial step of the synthesis was to mix oxindoles **99** with a variety of aryl nitriles **100**. A catalytic amount of saturated aqueous NaOH solution was then added to the mixture, the mixture was irradiated in a microwave oven at 180ºC for four minutes and the reaction progress was monitored with TLC every minute. The pyrimido[4,5-*b*]indoles were synthesized in yields ranging from 69-85%.

Borovik et al. [20] reported synthesizing 2-substituted-4-phenyl-9*H*-pyrimido[4,5-*b*]indoles (**104a-d**) from reaction of 2-ethoxy(3-benzylidene) indolenine tetrafluoroborate with various amidines or guanidines (Scheme 13). Starting compound **102** (3-benzylideneindolin-2-one) was treated with 2-ethoxy(3-benzylidene)indolenine tetrafluoroborate **103**. Condensation of **103** with acetimidamide, benzimidamide, guanidine, and *N*-carbamimidoyl substituted benzenesulfonamides yielded the compounds **104a**, **104b**, **104c**, and **104d** respectively.

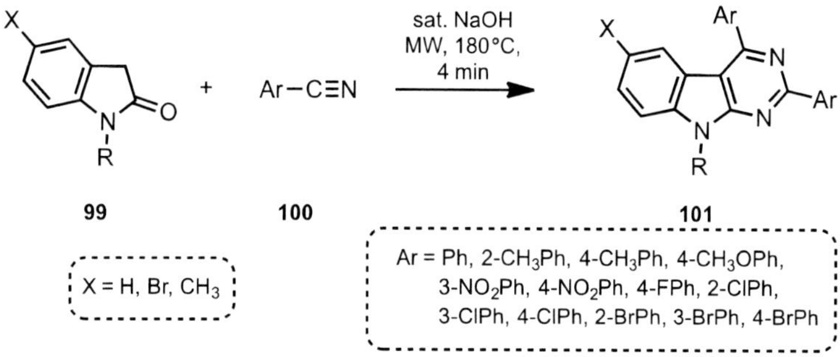

Scheme 12. Preparation of pyrimido[4,5-*b*]indoles from oxindoles.

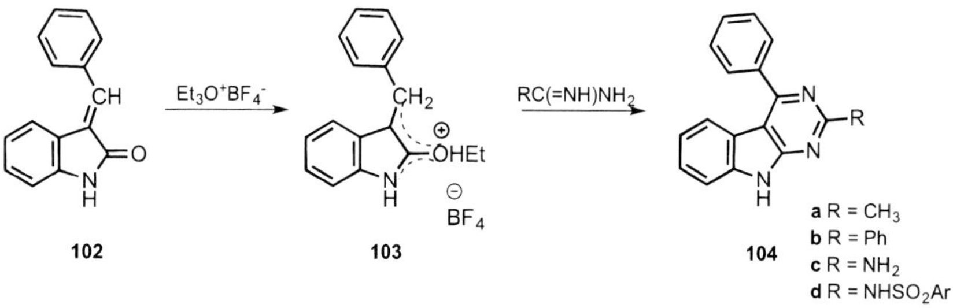

Scheme 13. Synthesis of from 3-benzylideneindolin-2-one to pyrimido[4,5-*b*]indoles.

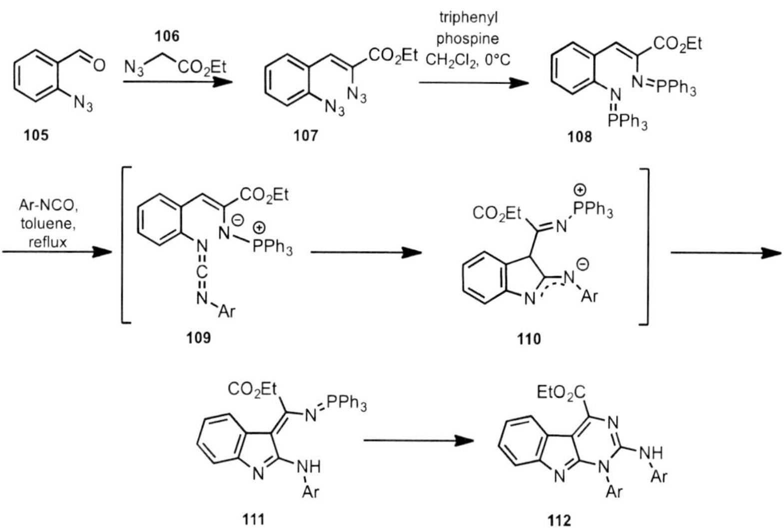

Scheme 14. Formation of pyrimido[4,5-*b*]indoles via aza-Wittig type reaction.

Molina et al. [21] reported the synthesis of pyrimido[4,5-*b*]indoles via aza Wittig type reaction (Scheme 14). The synthesis is initiated with the condensation of starting diazide *o*-azidobenzaldehyde **105** with ethylazidoacetate **106** which provided a diazide **107**. A Staudinger reaction between the diazide **107** and triphenylphosphine in dry dichloromethane at 0° C led to the bis(iminiphosphorane) **108**. Compound **108** reacts with aromatic iscyanates in toluene at reflux temperature to form the target molecules. To explain the conversion of **108** to **112**, it is proposed that the initial aza Wittig type reaction between the iminophosphorane group linked to the aromatic ring and the isocyanate supplies carbodiimide **109**, as a very reactive intermediate, that subsequently undergoes cyclization via nucleophilic attack of the β-carbon atom at the side chain leading to the formation of zwitterion intermediate **110**. Additional transformations of **110** will lead to the formation of indole **111**. Reacting bis(iminiphosphorane) **108** with aromatic isocyanates in toluene at reflux temperatures leads to initial formation of **111**. Reaction of **111** with a second mole of isocyanate forms a carbodiimide. Pyrimido-annelation occurs via intramolecular amination by nucleophilic attack of the exocyclic amino group on the central carbon atom of the carbodiimide moiety to afford target **112**.

Majumder et al. [22] developed a one-pot three-component synthesis of pyrimido[4,5-*b*]indoles in moderate to high yields (Scheme 15). This route is attractive as it offers cleaner reaction conditions to obtain pyrimido[4,5-*b*]indoles and precludes use of toxic organic solvent, expensive catalysts, and requirement of inert atmosphere. Five catalysts were initially screened: $InCl_3$, TsOH, CH_3COOH, L-proline, and $Yb(OTf)_3$ and yields varied from 60%, 25%, 65%, 55%, and 75%, respectively. The reaction in the presence of catalytic ytterbium triflate presented a satisfying result, 10 mol % of the catalyst was sufficient to produce the desired products. It was found that an increase to 20 mol% and 30 mol% of the catalyst did not improve the yields of the product.

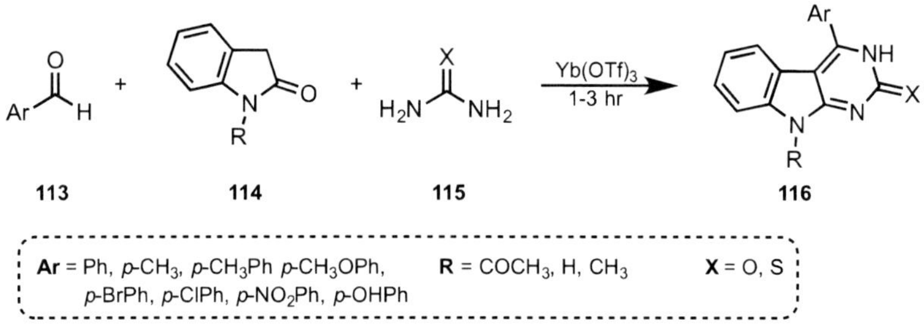

Scheme 15. One-pot synthesis of pyrimido[4,5-*b*]indoles via Yb(OTf)$_3$-catalysis in solvent-free condition.

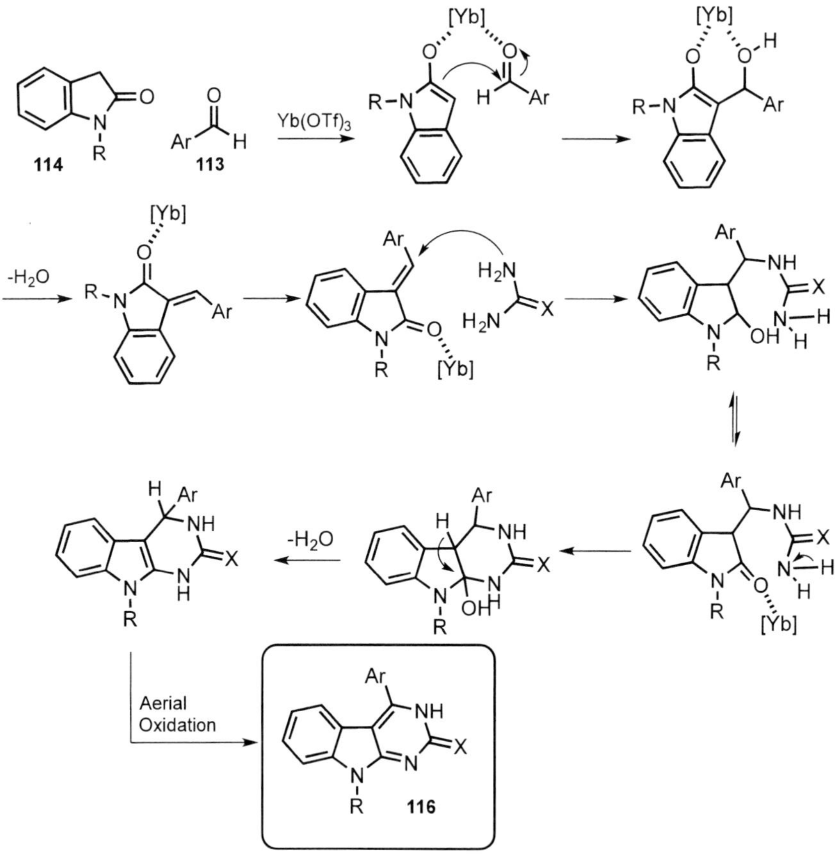

Scheme 16. Proposed mechanism for the transformation in Scheme 15.

Seventeen substrates were tested, all involving *ortho*- and *para*-substituted benzaldhydes, which provided pyrimido[4,5-*b*]indoles in yields ranging from 49 to 72%. Aryl aldehyde **113**, oxindole **114**, urea/thiourea **115**, and Lewis-acid catalyst ytterbium triflate $Yb(OTf)_3$ were all heated together in 100ºC for 1-3 hours. The proposed mechanism is depicted in Scheme 16 and indicates an initial Lewis acid-catalyzed cross-aldol condensation of the aryl aldehyde with oxindole. This is possibly followed by formation of a 3-substituted indole intermediate, and subsequent cyclocondensation to **116**.

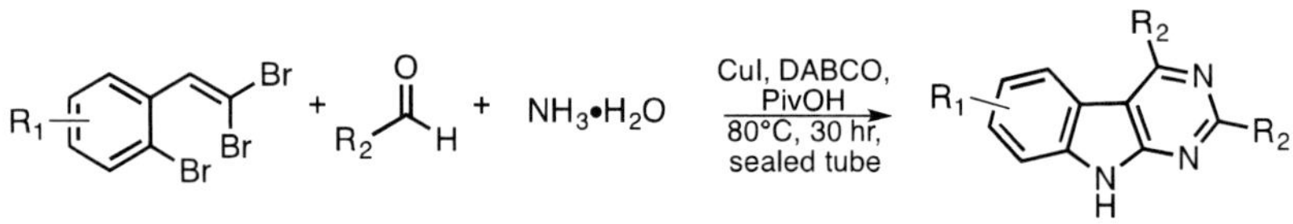

Scheme 17. Synthesis of 9*H*-pyrimido[4,5-*b*]indoles via copper-catalyzed one-pot multi-component cascade reactions.

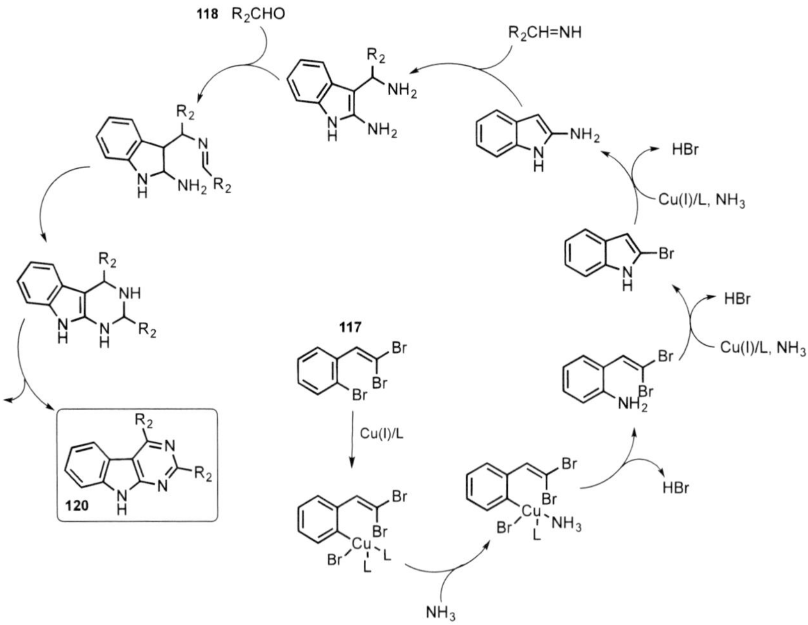

Scheme 18. Proposed mechanism for the transformation in Scheme 17.

Li et al. [23] developed a route for preparation of 9*H*-pyrimido[4,5-*b*]indoles that feature facile operation and the utilization of easily accessible starting materials (Scheme 17). The reactions presented in this study use less toxic reagents as well as avoiding harsh reaction conditions and multistep processes. It was found that a combination of DABCO, PivOH, and CuI showed an increase of product yield. PivOH has been frequently used in the past as an additive to facilitate Cu-catalyzed coupling reactions. The solvent of choice was found to be DMF for the reactions, whereas DMSO and NMP were less effective. Multiple catalysts were studied and it was found that CuI generated most product whereas CuBr and $Cu(OAc)_2$ resulted in decreased yields.

The scope for the synthesis of 9*H*-pyrimido[4,5-*b*]indoles were conducted on a range of 21 different substrates, which generated products ranging from 40-63% yields. The reactions began with the treatment of 1-bromo-2-(2,2-dibromomovinyl)benzene **117** (0.5 mmol), benzaldehyde **118** (1.5 mmol), and aqueous ammonia **119** with CuI (0.05 mmol), DABCO (0.1 mmol), and PivOH (0.5 mmol) in DMF at 80°C under air in a sealed tube for 30 hours. The proposed mechanism for the formation of 9*H*-pyrimido[4,5-*b*]indoles **120** is presented in Scheme 18.

CONCLUSION

The Pyrimido[4,5-*b*]indole core is a functionally versatile chemotype as higlighted by its presence in numerous pharmacologically active molecules. Recent advances for the chemical synthesis of these molecules are presented here. Pyrimido[4,5-*b*]indoles can be synthesized from a range of starting materials including substituted pyrimidines, pyrroles, and oxindoles among others. Sixteen synthetic strategies to access the pyrimido[4,5-*b*]indole system are described in this commentary. Recent modifications to previously reported reaction routes are also presented. The various synthetic protocols include but are not limited to catalysis, cyclizations, and multi-component one-pot synthesis. These reported methods aim to generate target molecules in high yields while reducing toxicity, making atom-economical and inexpensive use of catalysts, and achieving operational simplicity.

REFERENCES

[1] Zhou, Bing; Hu, Jiantao; Xu, Fuming; Chen, Zhuo; Bai, Longchuan; Fernandez-Salas, Ester; Lin, Mei; Liu, Liu; Yang, Chao-Yei; Zhao, Yujun; McEachern, Donna; Pryzbranowski, Sally; Wen, Bo; Sun, Duxin; Wang, Shaomeng. (2018). "Discovery of a Small-Molecule Degrader of Bromodomain and Extra-Terminal (BET) Proteins with Picomolar Cellular Potencies and Capable of Achieving Tumor Regression." *J. Med. Chem. 61:*462–481.

[2] Tichy, Michael; Pohl, Radek; Michal Tichy, Radek Pohl; Tloušt'ová, Eva; Weber, Jan; Bahador, Gina; Lee, Yu-Jen; Hocek, Michal. (2013). "Synthesis and biological activity of benzo-fused 7-deazaadenosine analogues. 5- and 6-substituted 4-amino- or 4-alkylpyrimido [4,5-*b*]indole ribonucleosides." *Bioorganic & Medicinal Chemistry* 21:5362–5372.

[3] Agarwal, Stuti; Agarwal, Hitesh; Zaware, Nilesh. (2017). "Pyrimido[4,5-b]indole derivatives and use thereof in the expansion of hematopoietic stem cells US2015011543 (a1): a patent evaluation." *Expert Opin Ther Pat.* 27:1-5.

[4] Gangjee, Aleem; Zaware, Nilesh; Raghavan, Sudhir; Disch, Bryan C. Disch; Thorpe, Jessica E.; Bastian, Anja; Ihnat, Michael A. (2013). "Synthesis and biological activity of 5-chloro-N^4-substitutedphenyl-9*H*-pyrimido[4,5-*b*]indole-2,4-diamines as vascular endothelial growth factor receptor-2 inhibitors and antiangiogenic agents." *Bioorganic & Medicinal Chemistry* 21:1857–1864.

[5] Gangjee, Aleem; Zaware, Nilesh; Devambatla, Ravi Kumar Vyas; Raghavan, Sudhir; Westbrook, Cara D.; Dybdal-Hargreaves, Nicholas F.; Hamel, Ernest; Mooberry, Susan L. (2013). "Synthesis of N^4-(substituted phenyl)-N^4-alkyl/desalkyl-9*H*-pyrimido[4,5-*b*]indole-2,4 -diamines and identification of new microtubule disrupting com-

pounds that are effective against multidrug resistant Cells." *Bioorganic & Medicinal Chemistry* 21:891–902

[6] Zaware, Nilesh; Sharma, Hitesh; Yang, Jie; Devambatla, Ravi Kumar Vyas; Queener, Sherry F.; Anderson, Karen S.; Gangjee, Aleem. (2013). "Discovery of Potent and Selective Inhibitors of *Toxoplasma gondii* Thymidylate Synthase for Opportunistic Infections." *ACS Medicinal Chemistry Letters* 4:1148-1151.

[7] Zhang, Yue-Mei; Razler, Thomas; Jackson, Paul F. (2002). "Synthesis of pyrimido[4,5-b]indoles and benzo[4,5]furo[2,3-d]pyrimidines via palladium-catalyzed intramolecular arylation." *Tetrahedron Letters* 43:8236-8239.

[8] Dotzauer, Bernd; Troschuetz, Reinhard. (2004). "Synthesis of medicinally interesting 2,4-diamino-9H-pyrimido[4,5-*b*]indol-6-ols via extension of the Nenitzescu reaction." *Synlett* 1039-1043.

[9] Wright, G. E. (1976). "9H-Pyrimido[4,5-*b*]indole-2,4-diones. The acid-catalyzed cyclization of 6-(phenylhydrazino)uracils." *Journal of Heterocyclic Chemistry* 13:539-44.

[10] Lapachev, Vyacheslav. V.; Stadlbauer, Wolfgang; Kappe, Thomas. (1988). "4-Aminouracils and pyrimido[4,5-*b*]indolediones from 4-azidouracils." *Monatsh. Chem.* 119:97-102.

[11] Mauragis, Michael A.; Veley, Michael F.; Lipton, Michael. F. (1997). "Evaluation and Rapid Scale-Up of the Synthesis of the Pyrrolopyrimidines U-101033E and U-104067F." *Org. Process Res. Dev.* 1: 39-44.

[12] Traxler, Peter M.; Furet, Pascal; Mett, Helmut; Buchdunger, Elisabeth; Meyer, Thomas; Lydron, Nicholas. (1996). "4(Phenylamino)pyrrolopyrimidines: Potent and Selective, ATP Site Directed Inhibitors of the EGF-Receptor Protein Tyrosine Kinase." *J. Med. Chem.* 39:2285-2292.

[13] Showalter, H.D. Hollis; Bridges, Alexander J.; Zhou, Hairong; Sercel, Anthony D.; McMichael, Amy; Fry, David W. (1999). Tyrosine Kinase Inhibitors. *J. of Med. Chem.* 42 (26), pp 5464–5474.

[14] Venugopalan, B.; Desai, P. D.; De Souza, N. J. (1988). "Synthesis of 6,7-dimethoxypyrimido[4,5-*b*]indoles as potential antihypertensive agents." *Journal of Heterocyclic Chemistry* 25:1633-1639.

[15] Müller, Christa E.; Geis, Uli; Grahner, Bettina; Lanzner, Wolfgang; Eger, Kurt. (1996). "Chiral Pyrrolo[2,3-d]pyrimidine and Pyrimido[4,5-b]indole Derivatives: Structure-Activity Relationships of Potent, Highly Stereoselective A1-Adenosine Receptor Antagonists." *Journal of Medicinal Chemistry* 39:2482-2491.

[16] Kapti, Tolga; Dengiz, Cagatay; Balci, Metin. (2016). "The Chemistry of Ethyl 3-(2-Ethoxy-2-oxoethyl)-1*H*-indole-2-carboxylate: Synthesis of Pyrimido[4,5-*b*]indoles and Diethyl 4-Hydroxy-quinoline-2,3-dicarboxylate via Intramolecular Cyclizations." *Synthesis* 49:1898-1904.

[17] Gangjee, Aleem; Zaware, Nilesh; Raghavan, Sudhir; Inhat, Michael; Shenoy, Satyendra; Kisliuk, Roy L. (2010). "Single Agents with Designed Combination Chemotherapy Potential: Synthesis and Evaluation of Substituted Pyrimido[4,5-*b*]indoles as Receptor Tyrosine Kinase and Thymidylate Synthase Inhibitors and as Antitumor agents." *J. Med. Chem.* 53:1563-1578.

[18] Zaware, Nilesh; Kisliuk, Roy; Bastian, Anja; Ihnat, Michael A.; Gangjee, Aleem. (2017). "Synthesis and evaluation of 5-(arylthio)-9H-pyrimido[4,5-b]indole-2,4-diamines as receptor tyrosine kinase and thymidylate synthase inhibitors and as antitumor agents." *Bioorganic & Medicinal Chemistry Letters* 27:1602-1607.

[19] Adib, Mehdi; Mohammadi, Bagher; Bijanzadeh, Hamid Reza. (2008). "A novel and simple synthesis of 9H-pyrimido[4,5-b]indoles under microwave irradiation and solvent-free conditions." *Synlett* 177-180.

[20] Borovik, V.P.; Shkurko, O. P. (2002). "Synthesis of functional 2-substituted 4-phenyl-9H-pyrimido[4,5-b]indoles." *Russ. Chem. Bull.* 51:2129-2133.

[21] Molina, P.; Arques, A; Alias, A.; Vinader, M. V. (1991). "Iminiophosphrane-mediated synthesis of functionalized indoles and 1,3-benzodiazepines." *Tetrahedron Lett.* 32:4401-4.

[22] Majumder, Swarup; Bhuyan, Pulak J. (2014). "One-pot multi-component synthesis of pyrimido[4,5-*b*]indoles in solvent-free condition." *J. Iran Chem Soc* 11:993-996.

[23] Li, Bin; Guo, Shenghai; Zhang, Ju; Zhang, Xinying; Fan, Xuesen. (2015). "Selective Access to 3-Cyano-1*H*-indoles, 9*H*-Pyrimido[4,5-*b*]indoles, or 9H-Pyrido[2,3-*b*]indoles through Copper-Catalyzed One-Pot Multicomponent Cascade Reactions." *J. Org. Chem.* 80:5444-5456.

Biographical Sketches

Claudia C. Kim

Affiliation: Icahn School of Medicine at Mount Sinai

Education:

William Paterson University of New Jersey	B.S.	Chemistry	05/2016
Icahn School of Medicine at Mount Sinai	M.S.	Biomedical Sciences	06/2020

Research and Professional Experience:

2014 – 2017 Organic Chemistry Lab, Yalan Xing PhD, William Paterson University of New Jersey

2018 – Structural and Chemical Epigenomics, Ming-Ming Zhou, Icahn School of Medicine at Mount Sinai in New York, NY

Honors:

2016 Excellence in Research Award, Class of 2016, William Paterson University

2016 McCallum Award for Outstanding Contribution in the Chemistry Department and Research, William Paterson University

2016 Center for Research (CrF) Funding for Research

Publications from the Last 3 Years:

Catano, B., Lee, J., Kim, C., Farrell, D., Peterson, Jeffrey L., Xing, Y. (2015) Iron(III) catalyzed halo-functionalization of alkynes, *Tetrahedron Letters*, 56:4124.

Jiang, L., Gao, T., Li, Z., Sun, S., Kim, C., Huang, C., Guo, H., Wang, J., Xing, Y. (2016) Highly regioselective dipolar cycloadditions of nitrile oxides with α,β-acetylenic aldehydes, *Tetrahedron Letters*, 57:712.

Ming-Ming Zhou

Affiliation: Dr. Harold and Golden Lamport Professor and Chairman, Department of Pharmacological Sciences, Icahn School of Medicine at Mount Sinai

Education:

East China University of Science & Technology, PRC	B.E.	Chemical Engineering	1984
Michigan Technological University	M.S.	Chemistry	1988
Purdue University	Ph.D	Chemistry	1993
Abbott Laboratories, Chicago, IL	Postdoc	Structural Biology/NMR	1993-1996

Research and Professional Experience:

1993 – 1996 – Postdoctoral Fellow, Abbot Laboratories, Chicago, IL (Advisor: Stephen W. Fesik)

1997 – 2000 – Assistant Professor, Department of Physiology & Biophysics, Mount Sinai School of Medicine (MSSM), New York, NY

2001 – 2004 – Associate Professor, Department of Physiology & Biophysics, MSSM

2004 – 2005 – Professor, Department of Structural & Chemical Biology (formerly Physiology & Biophysics), MSSM

2005 – 2016 – Chairman, Department of Structural & Chemical Biology, Icahn School of Medicine at Mount Sinai (ISMMS, formerly MSSM, name change in 2012)
2009 – Co-Director, Experimental Therapeutics Institute, ISMMS
2005 – Dr. Harold and Golden Lamport Professor, Department of Pharmacological Sciences (*formerly Structural & Chemical Biology)*; Professor, Department of Oncological Sciences, ISMMS
2016 – Chairman, Department of Pharmacological Sciences, ISMMS (*merger of Dept. of Structural & Chemical Biology, and Dept. of Pharmacology & Systems Therapeutics in 2016*)

Professional Appointments:

Editorial Boards

2008 – 2013 – Associate Editor, *Cancer Research*
2009 – Editorial Board, *Journal of Molecular Cell Biology*
2010 – Editorial Board, *ACS Medicinal Chemistry Letters*
2010 – *Faculty of 1000* on *"Structure, and Transcription and Translation"*
2012 – Editorial Board, *Journal of Cancer Immunology*
2012 – 2015 – Editorial Board, *Molecular Cancer Therapeutics* (ACS)
2014 – 2017 – Editorial Board, *World Journal of Biological Chemistry*

Scientific Advisory Boards

2006 – Scientific Advisory Board, New York Structural Biology Center
2007 – 2009 Co-Chair, 2009 Program Committee for 53rd Annual Meeting, Biophysical Society
2008 – 2011 International Science Advisory Board, Structural Genomics Consortium (Toronto & Oxford)
2010 – Thematic Meeting Committee, Biophysical Society
2010 – Translational Science Advisory Board, New York Academy of Sciences
2012 – 2013 Scientific Advisory Board, Coferon, Inc.
2013 – 2015 – Co-Organizer, 2015 FASEB Research Conference on *"HDACs, Sirtuins and Reversible Acetylation in Signaling and Disease"*, (Co-Organizer, with David Sinclair)

Board of Directors
2004 – Board of Directors, New York Structural Biology Center
2013 – Co-Founder and Director, MGZ Pharmaceutical, Inc.
2013 – Founder and Director, Parkside Scientific Inc.

Scientific Consulting
2001 – Research Institute of Molecular Pathology (MIP), Vienna, Austria
2003 – Editorial Consultant, *Natural Structural Biology*
2008 – Institute of Science and Technology Austria
2010 – Institute of Structural Biology and Drug Discovery, Virginia Commonwealth University
2011, 2012 – Active Motif, Inc.
2012, 2013 – Coferon, Inc.
2013 – Founder and Chief Scientist, Parkside

Honors:
1999 – 2001 – American Cancer Society Young Investigator Award
2003 – GlaxoSmithKline Drug Discovery and Development Award
2006 – Dr. Harold and Golden Lamport Professorship in Physiology and Biophysics
2009 – Elected to the Academy of Sciences & Arts at Michigan Technological University
2012 – Elected to a Fellow of American Association for the Advancement of Science (AAAS)
2014 – Honorary Professor, Jilin University, China

Publications from the Last 3 Years:
Cheung, K.L., Lu, G., Sharma, R., Vincek, A., Zhang, R.H., Plotnikov, A.N., Zhang, F., Zhang, Q., Ju, Y., Hu, Y., Zhao, L., Han, X., Meslamani, J., Xu, F., Jaganathan, A., Shen, T., Zhu, H., Rusinova, E., Zeng, L., Zhou, J., Yang, J., Peng, L., Ohlmeyer, M., Walsh, M.J., Zhang, D.Y., Xiong, H.B., & Zhou, M.-M. (2017) Selective BET Bromodomain Inhibition Blocks Th17 Cell Differentiation and Ameliorates Colitis in Mice. *PNAS.* 114(11): 2952-2957.

Cheung KL, Zhang F, Jaganathan A, Sharma R, Zhang Q, Konuma T, Shen T, Lee JY, Ren C, Chen CH, Lu G, Olson MR, Zhang W, Kaplan MH, Littman DR, Walsh MJ, Xiong H, Zeng L, Zhou MM. (2017) Distinct Roles of Brd2 and Brd4 in Potentiating the Transcriptional Program for Th17 Cell Differentiation. *Mol Cell*, 65(6):1068-1080.

Martín-Martínez M, Pérez-Gordillo FL, Álvarez de la Rosa D, Rodríguez Y, Gerona-Navarro G, González-Muñiz R, Zhou MM. (2017) Modulating Mineralocorticoid Receptor with Non-steroidal Antagonists. New Opportunities for the Development of Potent and Selective Ligands without Off-Target Side Effects. *J Med Chem*, 60(7):2629-2650.

Li J, Rodriguez Y, Cheng C, Zeng L, Wong EY, Xu CY, Zhou MM, Xu PX. (2016) EYA1's Conformation Specificity in Dephosphorylating Phosphothreonine in Myc and Its Activity on Myc Stabilization in Breast Cancer. *Mol Cell Biol*, 37(1).

Meslamani J, Smith SG, Sanchez R, Zhou MM. (2016) Structural features and inhibitors of bromodomains. *Drug Discov Today Technol*, 19:3-15.

Zhang Q, Zeng L, Zhao C, Ju Y, Konuma T, Zhou MM. (2016) Structural Insights into Histone Crotonyl-Lysine Recognition by the AF9 YEATS Domain. *Structure*, S0969-2126.

Zhang Q, Zeng L, Shen C, Ju Y, Konuma T, Zhao C, Vakoc CR, Zhou MM (2016) Structural Mechanism of Transcriptional Regulator NSD3 Recognition by the ET Domain of BRD4. *Structure*, 24(7):1201-8.

Ren C, Smith SG, Yap KL, Li S, Li J, Mezei M, Vincek AS, Aguilo F, Walsh MJ, and Zhou MM (2016) Structure-Guided Discovery of Selective Antagonists for the Chromodomain of Polycomb Repressive Protein CBX7. *ACS Med. Chem. Lett.*, 7(6), 601–605.

Zhang G, Smith SG, Zhou MM. (2015) Discovery of Chemical Inhibitors of Human Bromodomains. *Chem Rev,*115(21):11625-68.

Bansal N, Petrie K, Christova R, Chung CY, Leibovitch BA, Howell L, Gil V, Sbirkov Y, Lee E, Wexler J, Ariztia EV, Sharma R, Zhu J, Bernstein E, Zhou MM, Zelent A, Farias E, Waxman S. (2015) Targeting the SIN3A-PF1 interaction inhibits epithelial to mesenchymal transition and maintenance of a stem cell phenotype in triple negative breast cancer. *Oncotarget*, 6(33):34087-105.

Kwon YJ, Petrie K, Leibovitch BA, Zeng L, Mezei M, Howell L, Gil V, Christova R, Bansal N, Yang S, Sharma R, Ariztia EV, Frankum J, Brough R, Sbirkov Y, Ashworth A, Lord CJ, Zelent A, Farias E, Zhou MM, Waxman S. (2015). Selective Inhibition of SIN3 Corepressor with Avermectins as a Novel Therapeutic Strategy in Triple-Negative Breast Cancer. *Mol Cancer Ther*, 14(8):1824-36.

Stratikopoulos, E.E., Dendy, M., Szabolcs, M., Khaykin, A.J., Lefebvre, C., Zhou, M.-M., & Parsons, R. (2015) Kinase and BET Inhibitors Together Clamp Inhibition of PI3K Signaling and Overcome Resistance to Therapy. *Cancer Cell,* 27, 837-851.

Ren C, Morohashi K, Plotnikov AN, Jakoncic J, Smith SG, Li J, Zeng L, Rodriguez Y, Stojanoff V, Walsh M, Zhou MM. (2015). Small-Molecule Modulators of Methyl-Lysine Binding for the CBX7 Chromodomain. *Chem Biol,* pii: S1074-5521.

Nilesh Zaware, PhD

Affiliation: Icahn School of Medicine at Mount Sinai

Education: Doctorate in Medicinal Chemistry

Research and Professional Experience/Professional Appointments:

Bharatividyapeeth's College of Pharmacy, Mumbai University	B.Pharm.	Pharmaceutical Science	05/2002
Duquesne University	Ph.D.	Medicinal Chemistry	08/2009
University of Pittsburgh	Postdoc	Organic Chemistry	08/2010
Duquesne University	Postdoc	Medicinal Chemistry	09/2011
Icahn School of Medicine at Mount Sinai	Postdoc	Pharmacological Sciences	06/2017

Icahn School of Medicine at Mount Sinai	Instructor	Pharmacological Sciences	07/2017-

Honors:

1. "Best Student Award" for overall performance in the 2001-2002 Bachelor of Pharmaceutical Sciences academic year (2001-2002) in Bharati Vidyapeeth's College of Pharmacy, Mumbai University, Navi Mumbai, Maharashtra, India
2. Ranked 5th in Mumbai University among 297 students in the final year of Bachelor of Pharmaceutical Sciences studies (2001-2002)

Publications from the Last 3 Years:

Original peer reviewed publications:

1. McClinch, Kimberly; Avelar, Rita A.; Callejas, David; Izadmehr, Sudeh; Wiredja, Danica; Perl, Abbey; Sangodkar, Jaya; Kastrinsky, David; Schlatzer, Daniela; Cooper, Maxwell; Kiselar, Janna; Stachnik, Agnes; Yao, Shen; Hoon, Divya; McQuaid, Dan; Zaware, Nilesh; Gong, Yixuan; Brautigan, David; Plymate, Stephen; Sprenger, Cynthia; Oh, William K.; Levine, Alice C.; Kirschenbaum, Alexander; Sfakianos, John; Sears, Rosalie; DiFeo, Analisa; Ioannou, Yiannis; Ohlmeyer, Michael; Narla, Goutham; Galsky, Matthew. Small Molecule Protein Phosphatase 2A Activators for the Treatment of Castration-Resistant Prostate Cancer. *Cancer Research* 2018 *78(8),* 2065-2080
2. Zaware, Nilesh; Zhou, Ming-Ming. Chemical modulators for epigenome reader domains as emerging epigenetic therapies for cancer and inflammation. *Current Opinion in Chemical Biology* 2017, *39*, 116-125.
3. Agarwal, Stuti; Agarwal, Hitesh; Zaware, Nilesh* Pyrimidio[4,5-b]Indole Derivatives and Use thereof in the Expansion of Hematopoietic Stem Cells US2015011543 (A1): a patent evaluation. *Expert Opinion on Therapeutic Patents. Invited Manuscript.* 2017, *27(11)*, 1177-1181.
4. Sangodkar, Jaya; Perl, Abbey; Tohme, Rita; Kiselar, Janna; Kastrinsky, David B.; Zaware, Nilesh; Izadmehr, Sudeh; Mazhar, Sahar; Wiredja, Danica; O'Connor, Caitlin M.; Hoon, Divya; Dhawan, Neil S.;

Schlatzer, Daniela; Yao, Shen; Leonard, Daniel; Borczuk, Alain; Gokulrangan, Giridharan; Wang, Lifu; Svenson, Elena; Farrington, Caroline C.; Yuan, Eric; Avelar, Rita A.; Stachnik, Agnes; Smith, Blake; Gidwani, Vickram; Giannini, Heather M.; McQuaid, Daniel; McClinch, Kimberly; Wang, Zhizhi; Levine, Alice C.; Sears, Rosalie C.; Chen, Edward; Duan, Qiaonan; Datt, Manish; Haider, Shozeb; Ma'ayan, Avi; DiFeo, Analisa; Sharma, Neelesh; Galsky, Matthew; Brautigan, David L.; Ioannou, Yiannis; Xu, Wenqing; Chance, Mark; Ohlmeyer, Michael; Narla, Goutham. Activation of tumor suppressor protein PP2A inhibits KRAS-driven tumor growth. *The Journal of Clinical Investigation* 2017, *127(6)*, 2081-2090.

5. Devambatla, Ravi Kumar Vyas; Li, Wei; Zaware, Nilesh; Choudhary, Shruti; Hamel, Ernest; Mooberry, Susan; Gangjee, Aleem. Design, Synthesis, and Structure–Activity Relationships of Pyrimido[4,5-*b*]indole-4-amines as Microtubule Depolymerizing Agents that are Effective against Multidrug Resistant Cells. *Bioorg. Med. Chem. Lett.* 2017, *27(15)*, 3423-3430.
6. Bastian, Anja; Matsuzaki, Satoshi; Humphries, Kenneth M.; Pharaoh, Gavin, A.; Doshi, Arpit; Zaware, Nilesh; Gangjee, Aleem; Ihnat, Michael. AG311, a small molecule inhibitor of complex I and hypoxia-induced Q4 HIF-1α stabilization. *Cancer Letters* 2017, *388*, 149-157.
7. Zaware, Nilesh; Kisluik, Roy; Bastian, Anja; Ihnat, Michael A; Gangjee, Aleem. Synthesis and Evaluation of 5-(Arylthio)-9*H*-pyrimido[4,5-*b*]indole-2,4-diamines as Receptor Tyrosine Kinase and Thymidylate Synthase Inhibitors and as Antitumor Agents. *Bioorg. Med. Chem.* 2017, *27(7)*, 1602-1607.

Granted patents:

1. Title of Invention: *"Cyclic Vinylogous Amides as Bromodomain Inhibitors."*
   Names of Inventors: Zhou, Ming-Ming; Ohlmeyer, Michael; Vincek, Adam; Zaware, Nilesh. US2016/02000666 A1, International publication date: 2016-07-14.

Patent applications:

1. Title of Invention: *"Cyclic Vinylogous Amides as Bromodomain Inhibitors."*
   Names of Inventors: Zhou, Ming-Ming; Ohlmeyer, Michael; Vincek, Adam; Zaware, Nilesh. PCT Int. Appl. (2015), WO 2015031824 A1, International publication date: 2015-03-05. – 3 citations.
2. Title of Invention: *"Stereospecific Process for 3-Heterocyclyl Cyclo Aliphatic 1,2-Diols"* Name of Inventors: Michael Ohlmeyer, Nilesh Zaware. PCT Int. Appl. (2017), WO 2017024229 A1, International publication date: 2017-02-09.
3. Title of Invention: *"Heterocyclic Constrained Tricyclic Sulfonamides as Anti-Cancer Agents"* Name of Inventors: Michael Ohlmeyer, Nilesh Zaware. PCT Int. Appl. (2017), WO 2017044567 A1, International publication date: 2017-03-16.
4. Title of Invention: *"Heterocyclic Constrained Tricyclic Sulfonamides as Anti-Cancer Agents"* Name of Inventors: Michael Ohlmeyer, Nilesh Zaware. PCT Int. Appl. (2017), WO 2017044569 A1, International publication date: 2017-03-16.
5. Title of Invention: *"Tricyclic Sultam Sulfonamides as Anticancer and Neuroprotective Agents"* Name of Inventors: Michael Ohlmeyer, David Kastrinsky, Nilesh Zaware. PCT Int. Appl. (2017), WO 2017044571 A1, International publication date: 2017-03-16.

Bibliography

2-phenylindole-3-aldehyde and certain of its derivatives

LCCN	a 46005972
Type of material	Book
Personal name	Blume, Roe Calvin, 1916- [from old catalog]
Main title	2-phenylindole-3-aldehyde and certain of its derivatives
Published/Created	[n.p., 1946]
Description	185-188, 255-258 p. 26 cm.
LC classification	QD401 .B65
Related names	Lindwall, Harry Gustave, 1902- [from old catalog] joint author.
Subjects	Indole. Aldehydes.

3-indole acetaldehyde as a growth hormone in higher plants

LCCN	46019551
Type of material	Book
Personal name	Larsen, Poul, 1909-
Main title	3-indole acetaldehyde as a growth hormone in higher plants / Poul Larsen.

Published/Created	København: Munksgaard, 1944.
Description	132 p.: ill.; 24 cm.
LC classification	QK731 .L32
Subjects	Plant hormones. Acetaldehyde. Plants--Effect of light on.
Notes	Issued also as thesis (Copenhagen). Translation of: [Beta]-indolyl-acetaldehyd som væksthormon hos højere Planter. "Translated from Danish by Niels Haislund."--Verso of t.p. "Reprinted from Dansk botanisk arkiv (Res botanicae danicae) vol. 11, no. 9." Bibliography: p. [129]-132.

A biochemical study of Pacific coast salmon with particular reference to the formation of indol and skatol during decomposition,

LCCN	23014802
Type of material	Book
Personal name	Clough, Ray William, 1885-
Main title	A biochemical study of Pacific coast salmon with particular reference to the formation of indol and skatol during decomposition, by Ray William Clough ...
Published/Created	Seattle, University Press, University of Washington, 1922.
Description	195-272 p. diagrs. 26 cm.
LC classification	TX612.S4 C6
Subjects	Salmon. Indole. Skatol.
Notes	Cover-title.

Published also in Washington (State) University. Puget Sound Biological Station. Publications. v. 3, no. 67.
Bibliography: p. 268-272.
Dissertation note Thesis (PH. D.)--University of Washington.

An evidence-based approach to dietary phytochemicals

LCCN 2006017925
Type of material Book
Personal name Higdon, Jane.
Main title An evidence-based approach to dietary phytochemicals / Jane Higdon.
Published/Created New York: Thieme Medical Publishers, c2007.
Description xvi, 238 p.: ill.; 25 cm.
Links Table of contents only http://www.loc.gov/catdir/toc/ecip0614/2006017925.html
ISBN 9781588904089 (alk. paper)
1588904083 (alk. paper)
9783131418418 (alk. paper)
3131418419 (alk. paper)
LC classification QP144.V44 H54 2007
Contents Fruits and vegetables -- Cruciferous vegetables -- Legumes -- Nuts -- Whole grains -- Coffee -- Tea -- Carotenoids -- Chlorophyll and chlorophyllin -- Curcumin -- Essential fatty acids (omega-3 and omega-6) -- Fiber -- Flavonoids -- Soy isoflavones -- Isothiocyanates -- Indole-3-carbinol -- Lignans -- Organosulfur compounds from garlic -- Phytosterols -- Resveratrol.
Subjects Phytochemicals--Physiological effect.
Evidence-based medicine.
Plant Extracts--therapeutic use.
Dietary Supplements.
Phytotherapy.

	Plants, Edible--chemistry.
Notes	Includes bibliographical references and index.
Series	Complementary medicine

Bad nature, or with Elvis in Mexico

LCCN	2009039623
Type of material	Book
Personal name	Marías, Javier.
Uniform title	Mala indole. English
Main title	Bad nature, or with Elvis in Mexico / Javíer Marias; translated by Esther Allen.
Published/Created	New York: New Directions, 2010.
Description	57 p.; 18 cm.
ISBN	9780811218580 (pbk.: acid-free paper) 0811218589
LC classification	PQ6663.A7218 M3513 2010
Related names	Allen, Esther, 1962-
Subjects	Presley, Elvis, 1935-1977--Fiction.
Notes	"A New Directions Pearl."

Biosynthesis and molecular genetics of fungal secondary metabolites

LCCN	2014946216
Type of material	Book
Main title	Biosynthesis and molecular genetics of fungal secondary metabolites / Juan-Francisco Martin, Carlos García-Estrada, Susanne Zeilinger, editors.
Published/Produced	New York: Springer, [2014]
Description	volumes: illustrations (some color); 24 cm.
Links	Table of contents http://catdir.loc.gov/catdir/enhancements/fy1412/2014946216-t.html Contributor biographical information http://catdir.loc.gov/catdir/enhancements/fy1412/2014946216-b.html

	Publisher description http://catdir.loc.gov/catdir/enhancements/fy1412/2014946216-d.html
ISBN	9781493911905 (v. 1: alk. paper) 1493911902 (v. 1: alk. paper) 9781493911912 (v. 1: eBook) 1493911910 (v. 1: eBook) 9781493925308 (v. 2: alk. paper) 149392530X (v. 2: alk. paper) (v. 2: eBook)
LC classification	QK601 .B56 2014
Related names	Martín, Juan Francisco, editor. García-Estrada, Carlos, editor. Zeilinger, Susanne, editor.
Summary	This volume describes the more relevant secondary metabolites of different fungi with current information on their biosynthesis and molecular genetics. Bolstered with color illustrations and photographs, the book describes the possible application of molecular genetics to directed strain improvement in great detail. The needs for future developments in this field are also discussed at length. Written by authorities in the field, Biosynthesis and Molecular Genetics of Fungal Secondary Metabolites provides a cutting-edge perspective on fungal secondary metabolism and underlying genetics and is a valuable resource for scientists, researchers, and educators in the field of fungal biology. -- Source other than Library of Congress.
Contents	Valuable Secondary Metabolites from Fungi -- Penicillins -- Cephalosporins -- Cyclosporines: Biosynthesis and Beyond -- Aflatoxin Biosynthesis: Regulation and Subcellular Localization -- Roquefortine C and Related

	Prenylated Indole Alkaloids -- Ochratoxin A and Related Mycotoxins -- Carotenoids -- Astaxanthin and Related Xanthophylls -- Gibberellins and the Red Pigments Bikaverin and Fusarubin -- Fusarins and Fusaric Acid in Fusaria -- Lovastatin, Compactin, and Related Anticholesterolemic Agents -- Meroterpenoids -- Ergot Alkaloids -- Fungal NRPS-dependent Siderophores: from Function to Prediction.
Subjects	Fungal metabolites. Biosynthesis. Metabolism, Secondary. Fungi--metabolism Fungi--genetics Biological Products Metabolomics Cell Biology. Microbiology. Plant Genetics & Genomics. Plant Biochemistry. Science--Life Sciences--Anatomy & Physiology. Biosynthesis. Fungal metabolites. Metabolism, Secondary.
Notes	Includes bibliographical references and index.
Series	Fungal Biology, 2198-7777 Fungal biology (Springer (Firm)), 2198-7777

Biosynthesis of indole alkaloids

LCCN	81022591
Type of material	Book
Personal name	Rahman, Atta-ur-, 1942-
Main title	Biosynthesis of indole alkaloids / by Atta-ur-Rahman and Anwer Basha.

Published/Created	Oxford [Oxfordshire]: Clarendon Press; New York: Oxford University Press, 1983.
Description	270 p.: ill.; 24 cm.
ISBN	0198556101:
LC classification	QP801.I45 R256 1983
Related names	Basha, Anwer, 1947-
Subjects	Indole alkaloids--Synthesis.
Notes	Includes bibliographical references and index.
Series	The International series of monographs on chemistry; 7 Oxford science publications

Biotechnology of hairy root systems

LCCN	2013943946
Type of material	Book
Main title	Biotechnology of hairy root systems / Pauline M. Doran, editor; with contributions by Z. Al-Shalabi ... [and 26 others].
Published/Produced	Heidelberg; New York: Springer, [2013]
Description	vii, 159 pages: illustrations (some color); 24 cm
ISBN	9783642390180 (alk. paper) 3642390188 (alk. paper)
LC classification	QK644 .B56 2013
Related names	Doran, Pauline M., editor.
Contents	Recent advances in the understanding of Agrobacterium rhizogenes-derived genes and their effects on stress resistance and plant metabolism / Victor P. Bulgakov and Yuri N. Shkryl -- Perspectives of the metabolic engineering of terpenoid indole alkaloids in Catharanthus roseus hairy roots / Le Zhao, Guy W. Sander and Jacqueline V. Shanks -- Effective elicitors and process strategies for enhancement of secondary metabolite production in hairy root

	cultures / Jian Wen Wang and Jian Yong Wu -- Hairy root culture: bioreactor design and process intensification / Amanda R. Stiles and Chun-Zhao Liu -- Hairy roots as a vaccine production and delivery system / Marina Skarjinskaia, Karen Ruby and Adriana Araujo -- Metal uptake and nanoparticle synthesis in hairy root cultures / Zahwa Al-Shalabi and Pauline M. Doran.
Subjects	Roots (Botany)--Research. Root hairs. Plant Roots--genetics. Bioreactors. Metabolic Engineering--methods. Plant Roots--growth & development. Plant Roots--metabolism. Root hairs. Roots (Botany)--Research.
Notes	Includes bibliographical references and index.
Series	Advances in biochemical engineering/ biotechnology, 0724-6145; 134

C-H activation

LCCN	2010927936
Type of material	Book
Main title	C-H activation / volume editors: Jin-Quan Yu and Zhangjie Shi; with contributions by L. Ackermann ... [et al.].
Published/Created	Heidelberg; New York: Springer, c2010.
Description	xx, 384 p.: ill.; 24 cm.
ISBN	9783642123559 (alk. paper) 3642123554 (alk. paper) 9783642123566 (e-ISBN) 3642123562 (e-ISBN)
LC classification	QD411 .C15 2010

QD1 .F58 v.292

Related names Yu, Jin-Quan.
Ackermann, Lutz.
Shi, Zhangjie, 1974-

Contents Synthesis in the key of Catellani: Norbornene-mediated ortho C-H functionalization / Andrew Martins, Brian Mariampillai, and Mark Lautens -- Mechanistic considerations in the development and use of azine, diazine and azole N-oxides in palladium-catalyzed direct arylation / Keith Fagnou -- Palladium and copper catalysis in regioselective, intermolecular coupling of C-H and C-Hal bonds / Olafs Daugulis -- Pd-catalyzed C-H bond functionalization on the indole and pyrrole nucleus / Elizabeth M. Beck and Matthew J. Gaunt -- Remote C-H activation via through-space palladium and rhodium migrations / Feng Shi and Richard C. Larock -- Palladium-catalyzed aryl-aryl bond formation through double C-H activation / Shu-Li You and Ji-Bao Xia -- Palladium-catalyzed allylic C-H bond functionalization of olefins / Guosheng Liu and Yichen Wu -- Ruthenium-catalyzed direct arylations through C-H bond cleavages / Lutz Ackermann and Rubén Vicente -- Rhodium-catalyzed C-H bond arylation of arenes / Jean Bouffard and Kenichiro Itami -- Cross-dehydrogenative coupling reactions of sp3-hybridized C-H bonds / Woo-Jin Yoo and Chao-Jun Li -- Functionalization of carbon-hydrogen bonds through transition metal carbenoid insertion / Huw M.L. Davies and Allison R. Dick -- Metal-catalyzed oxidations of C-H to C-N bonds / David N. Zalatan and J. Du Bois.

Subjects	Organometallic compounds. Activation (Chemistry) Organometallic chemistry. Carbon--chemistry. Hydrogen--chemistry.
Notes	Includes bibliographical references and index.
Additional formats	Also available online to subscrbing institutions.
Series	Topics in current chemistry, 0340-1022; v. 292

Chemical studies on some indole derivatives.

LCCN	74193386
Type of material	Book
Personal name	Bergman, Jan, 1941-
Main title	Chemical studies on some indole derivatives.
Published/Created	Stockholm, 1971.
Description	21 l. illus. 30 cm.
LC classification	QD401 .B487
Subjects	Indole.
Notes	Includes bibliographical references.
Dissertation note	Avhandling--Stockholm.

Chloroplast biotechnology: methods and protocols

LCCN	2014931084
Type of material	Book
Main title	Chloroplast biotechnology: methods and protocols / edited by Pal Maliga.
Published/Produced	New York: Humana Press, [2014]
Description	xv, 452 pages: illustrations; 27 cm.
ISBN	9781627039949 (hardback: alk. paper) 1627039945 (hardback: alk. paper) 1627039953 9781627039956
LC classification	QK725 .C474 2014
Related names	Maliga, P., editor.

Summary In Chloroplast Biotechnology: Methods and Protocols, expert researchers in the field detail many of the methods which are now commonly used in chloroplast molecular biology. Chapters focus on essential background information, applications in tobacco, and protocols for plastid transformation in crops, and Chlamydomonas and Bryophytes. Written in the highly successful Methods in Molecular Biology series format, chapters include introductions to their respective topics, lists of the necessary materials and reagents, step-by-step, readily reproducible laboratory protocols, and key tips on troubleshooting and avoidance of known pitfalls. Authoritative and practical, Chloroplast Biotechnology: Methods and Protocols seeks to aid scientists who study chloroplast molecular biology as well as those interested in applications in agriculture, industrial biotechnology, and healthcare.-- Source other that the Library of Congress.

Contents The Plastid Genomes of Flowering Plants -- Next Generation Technologies to Determine Plastid Genome Sequences -- Plastid Gene Transcription: Promoters and RNA Polymerases -- Plastid mRNA Translation -- Engineering Chloroplasts for High-Level Foreign Protein Expression -- Excision of Plastid Marker Genes Using Directly Repeated DNA Sequences -- Fluorescent Labeling and Confocal Microscopic Imaging of Chloroplasts and Non-Green Plastids -- Plastid Transformation in Nicotiana tabacum and Nicotiana sylvestris by Biolistic DNA Delivery to Leaves -- Nicotiana tabacum PEG-mediated

Plastid Transformation -- Plastid Transformation of Tobacco Suspension Cell Cultures -- Tryptophan and Indole Analog Mediated Plastid Transformation -- Plastid Marker Gene Excision in Greenhouse-Grown Tobacco by Agrobacterium-Delivered Cre Recombinase -- Determination of the Half-Life of Chloroplast Transcripts in Tobacco Leaves -- Quantification of Organellar DNA and RNA Using Real-Time PCR -- Plastid Transformation for Rubisco Engineering and Protocols for Assessing Expression -- Plastid Transformation in Tomato -- Stable Plastid Transformation of Petunia -- Plastid Transformation in Potato -- Solanum tuberosum -- Plastid Transformation in Eggplant -- Plastid Transformation in Lettuce (Lactuca sativa L.) by Polyethylene Glycol Treatment of Protoplasts -- Plastid Transformation in Lettuce (Lactuca sativa L) by Biolistic DNA Delivery -- Plastid Transformation in Soybean -- Plastid Transformation in Cabbage (Brassica oleracea L. var. capitata L.) by the Biolistic Process -- Plastid Transformation in Sugar Beet Beta vulgaris -- Integration and Expression of gfp in the Plastid of Medicago sativa L. -- Rapid Screening for the Robust Expression of Recombinant Proteins in Algal Plastids -- A Simple, Low Cost Method for Chloroplast Transformation of the Green Alga Chlamydomonas reinhardtii -- Tools for Regulated Gene Expression in the Chloroplast of Chlamydomonas -- Plastid Transformation in Physcomitrella patens -- Plastid Transformation of Sporelings and Suspension-cultured Cells from the Liverwort Marchantia polymorpha L.

Subjects	Chloroplasts.
	Biotechnology.
	Chloroplasts--metabolism--Laboratory Manuals.
	Chloroplasts--genetics--Laboratory Manuals.
	Genetic Engineering--methods--Laboratory Manuals.
	Plastids--genetics--Laboratory Manuals.
	Plastids--metabolism--Laboratory Manuals.
	Transformation, Genetic--Laboratory Manuals.
	Biotechnology.
	Chloroplasts.
Form/Genre	Laboratory Manuals.
Notes	Includes bibliographical references and index.
Series	Methods in molecular biology (Clifton, N.J.); v. 1132. 1064-3745
	Springer protocols (Series) 1948-3448

Gall formation by Phytomonas tumefaciens extract and indole-3-acetic acid in cultures of tomato roots

LCCN	a 43000662
Type of material	Book
Personal name	Friedman, B. A. (Bernard A.), 1909-
Main title	Gall formation by Phytomonas tumefaciens extract and indole-3-acetic acid in cultures of tomato roots
Published/Created	[n. p.] [1942]
Description	762-772 p. incl. illus., tables. 27 cm.
LC classification	SB767 .F7
Related names	Francis, Thomas, 1900-1969, joint author.
Subjects	Crown-gall disease.
	Indoleacetic acid.
	Agrobacterium tumefaciens.
	Plant diseases.

Gardening indoors with soil & hydroponics

LCCN	2011282361
Type of material	Book
Personal name	Van Patten, George F.
Uniform title	Gardening indoors
Main title	Gardening indoors with soil & hydroponics / by George F. Van Patten.
Edition	5th ed.
Published/Created	[S.l.]: Van Patten Pub., c2008.
Description	x, 374 p.: chiefly col. ill.; 21 cm.
ISBN	1878823329 (pbk.) 9781878823328 (pbk.)
LC classification	SB419 .V366 2008
Variant title	Gardening indoors with soil and hydroponics
Summary	From back cover: A best seller since 1986, now expanded and completely rewritten with all new information! More than 670 full color photos, drawings, charts and graphs illustrate every detail of this authoritative easy-to-understand book. Numerous simple cultivation solutions make it appealing to novice gardeners. Seasoned gardeners are also able to find answers to all their questions. Discover how to achieve the biggest best yields even with limited space and a small budget. Learn how to grow while taking an exclusive photo tour of some of the most interesting indoor gardens and greenhouses in the world.
Contents	Introduction -- Credits -- History of Hydroponics -- Table of contents -- Internet -- Trade shows & expos -- 1: Horticulture -- Introduction -- Life cycle of annual vegetables and flowers -- Germination -- Seedling growth -- Vegetative growth -- Mother plants -- Taking cuttings --

Flowering -- 2: Seeds & Seedlings -- Introduction -- Seeds -- Germination -- Water -- Temperature -- Air (oxygen) -- Two popular germination techniques -- One: pre-soaking in water -- Two: direct seed -- Ordering seeds -- Storing seeds -- Seed pests -- Seedlings -- Buying seedlings -- 3: Vegetative Growth -- Introduction -- Cuttings and taking cuttings -- Mother plants -- Getting ready -- Precautions -- Rooting hormones -- Before taking cuttings -- Taking cuttings: step-by-step -- Cuttings from a flowering plant -- Storing cuttings -- Transplanting -- Transplanting: step-by-step -- Hardening off -- Pruning and bending -- Pruning -- Bending -- Air pruning roots -- Root pruning -- Chemical root pruning -- Stress -- 4: Flowering -- Introduction -- Long-day plants -- Hermaphrodites -- Seed crops -- Life cycle -- 5: Harvest -- Introduction -- Harvest -- 6: Indoor Gardens And Greenhouses -- Introduction -- Setting up the indoor garden: step-by-step -- Indoor garden checklist -- Greenhouses and cold frames -- Coverings -- Climate control -- 7: Light, Lamps & Electricity -- Introduction -- Light, spectrum, and photoperiod -- PAR and light spectrum -- Measuring light -- Light meters -- Photoperiod -- Intensity -- Inverse square law -- Lamp spacing -- Side lighting -- Rotating plants -- Plant spacing -- Reflective hoods -- Horizontal reflective hoods -- Vertical reflective hoods -- Air-cooled lamp fixtures -- No reflective hood -- Reflective hood study -- Reflective light -- More free growing light -- Light movers -- Setting up a light mover: step-by-step -- High intensity discharge (HID) lights -- HID ballasts -- HID bulbs -- Metal halide systems -- Construction & operation -- Lumen maintenance and life -- Metal halide ballasts -- Metal halide bulbs -- High pressure sodium systems -- Operation and construction -- Life and lumen maintenance -- HP sodium ballasts -- HP sodium bulbs -- Conversion bulbs -- HP sodium to metal halide -- Metal halide to HP sodium -- Mercury vapor lamps -- Fluorescents lamps -- Construction and operation -- Other lamps -- Incandescent lamps -- LP sodium lamps -- Tungsten halogen lamps -- Electricity & safety -- Electricity consumption -- Generators -- Timers -- Setting up the HID system: step-by-step -- 8: Soil And Containers -- Introduction -- PH -- Humates chelate -- Soil temperature -- Potting soil -- Mushroom compost -- Soilless mix -- Cutting and seedling -- Cubes and mixes -- Soil amendments -- Soil mixes -- Compost -- Growing medium disposal -- Growing medium problems -- Containers -- Drainage -- Container shape, size, and maintenance -- Green roots -- 9: Water & Nutrients -- Introduction -- Water quality -- Hard water -- Sodium chloride -- Osmosis -- Reverse osmosis -- Irritation -- Overwatering -- Underwatering -- Drip systems -- Misdiagnosed disorders -- Nutrients -- Macronutrients -- Nitrogen (N) mobile -- Phosphorus (P) mobile -- Potassium (K) mobile -- Secondary nutrients -- Magnesium (Mg) mobile -- Calcium (Ca) immobile -- Chelates -- Sulfur (S) immobile -- Micronutrients -- Zinc (Zn) mobile -- Manganese (Mn) immobile -- Iron (Fe) immobile -- Boron (B) immobile -- Chlorine (chloride) (CL) immobile -- Cobalt (CO) immobile -- Copper (Cu) immobile -- Molybdenum (Mb) immobile -- Silicon (SI) immobile -- Nickel (Ni) -- Sodium (Na) -- Fluoride (F) -- Other elements -- Fertilizers -- Chemical fertilizers -- Organic fertilizers -- Organic teas -- Mixing fertilizers -- Fertilizer application -- Foliar feeding -- Spreaders-stickers -- Additives -- Abscisic acid (ABA) -- Ascorbic acid (vitamin C) Aspirin -- Auxins -- Bacteria -- B 9 folic acid -- Cellulase -- Cytokinins -- Enzymes -- Ethylene gas -- Flower saver plus -- Fulvic acid -- Gibberellins -- Humic acid -- Hydrogen peroxide -- Indole 3 butyric acid (IBA) -- Isopentyl adenine (IPA) -- Rhizobuim -- Spray-n-grow -- Sugar -- Trichoderma (002/003) -- Zeatin -- 10: Hydroponic Gardening -- Introduction -- Hydroponic systems -- Ebb and Flow gardens -- Build your own Ebb & Flow system -- Air tables -- Deep water culture (DWC) -- Top-feed systems -- Top-feed buckets -- Multiple bucket top-feed -- Top-feed slabs: slabs in individual trays -- Tables of slabs -- Individual blocks -- Vertical top-feed systems -- Nutrient film technique (NFT) -- Bioponic culture -- Aeroponics -- Growing mediums -- Sterilizing -- PH -- EC, TDS, DS, CF, PPM -- EC meters -- Hydroponic nutrients -- Nutrient solutions -- Solution maintenance -- Hydro-organic -- Nutrient solution composition -- Reservoirs -- Reservoir temperature -- Irrigation -- Nutrient disorders -- 11: Air -- Introduction -- Air movement -- Stomata -- Circulation -- Ventilation -- Intake air -- Temperature -- Thermostats -- Humidity -- Measuring relative humidity -- CO 2 enrichment -- Measuring CO -- Producing CO 2 -

- CO 2 emitter systems -- CO 2 generator systems -- Other ways to make CO 2 -- Compost and organic growing mediums -- Fermentation -- Dry ice -- Baking soda and vinegar -- Negative ion generators -- Ozone generators -- Ozone damage -- Activated carbon filters -- Setting up the vent fan: step-by-step -- 12: Pests And Diseases -- Introduction -- Prevention -- Control -- Insect control -- Sprays and traps, chemical fungicides, insecticides, and miticides -- Spreader-sticker for pesticides -- Abamectin -- Bacillus thuringiensis (Bt) and other bacillus species -- Baking soda -- Bleach, laundry -- Bordeaux mixture -- Boric acid -- Bug bombs -- Copper -- Diatomaceous earth (DE) -- Homemade pest and disease sprays -- Neem -- Neem oil -- Nicotine and tobacco sprays -- Oil, horticultural -- Pyrethrum -- Synthetic pyrethroids -- Quassia -- Rotenone -- Ryania -- Sabadilla -- Seaweed -- Soap, insecticidal -- Sulfur -- Traps -- Water -- Biological controls -- Predators and parasites -- Spider mites and insects -- Spider mites -- Aphids -- Bees & wasps -- Beetle borers -- Caterpillars and loopers -- Leafhoppers -- Leaf miners -- Fungus gnats -- Mealybugs and scale -- Nematodes -- Root maggot -- Slugs and snails -- Thrips -- Whiteflies -- Fungi and diseases -- Prevention -- Control -- Gray mold (botrytis) -- Damping-off -- Downy mildew -- Blight -- Foliar spots and fungi -- Fusarium wilt -- Green algae -- Powdery mildew -- Root rot -- Pythium wilt/rot -- Sooty mold -- Verticilluim wilt -- Viruses -- Tomato problems images -- Spraying -- 13: Plant Selection Guide -- Ornamentals -- Orchids -- Vegetables --

	Appendix -- Conversion charts and tables -- Calculations for metric users -- Glossary -- Index -- Advertising.
Subjects	Indoor gardening. Hydroponics.
Notes	Edition statement appears on p. [4] of cover. Includes index.

Grandeza y decadencia de España.

LCCN	51025667
Type of material	Book
Personal name	Maura, Gabriel Maura Gamazo, duque de, 1879-
Main title	Grandeza y decadencia de España.
Published/Created	Madrid, Ediciones Ambos Mundos [1947?]
Description	278 p. 20cm.
LC classification	DP68 .M35
Subjects	Spain--History. Spain--Civilization.
Notes	"Se insertan en este volumen siete trabajos de índole varia, a saber: un discurso do recepción en la Real Academia Española, un prólogo, tres conferencias leídas en centros culturales, uno de ellos protugués, y dos artículos de crítica bibliográfica, publicados por el Boletín de la Real Academia de la Historia."
Series	His Obras completas, v. 1 Maura, Gabriel Maura Gamazo, duque de, 1879- Obras completas, v. 1.

Halogen interchange during the zinc halide induced Fischer indole reactions of acetophenone, 2,6-dihalophenylhydrazones.

LCCN	55037685
Type of material	Book
Personal name	Larson, Gerald W. [from old catalog]

Main title	Halogen interchange during the zinc halide induced Fischer indole reactions of acetophenone, 2,6-dihalophenylhydrazones.
Published/Created	Pittsburgh, 1953.
Description	viii, 78 l. diagrs., tables. 29 cm.
LC classification	QD501 .L28
Related titles	Fischer indole reaction.
Subjects	Acetophenone.
	Phenylhdrazones. [from old catalog]

Heterocyclic compounds with indole and carbazole systems

LCCN	53010760
Type of material	Book
Personal name	Sumpter, Ward C.
Main title	Heterocyclic compounds with indole and carbazole systems [by] Ward C. Sumpter [and] F.M. Miller.
Published/Created	New York, Interscience Publishers, 1954.
Description	307 p. illus. 24 cm.
LC classification	QD331 .S8
Related names	Miller, F. M. (Fred Mathias), 1905- joint author.
Subjects	Heterocyclic compounds.
	Indole.
	Carbazole.
Series	The Chemistry of heterocyclic compounds, v. 8

High resolution ultraviolet spectroscopic studies of indole derivatives and their clusters

LCCN	92956480
Type of material	Book
Personal name	Wu, Yenchune Ruth, 1960-
Main title	High resolution ultraviolet spectroscopic studies of indole derivatives and their clusters [microform] / by Yenchune Ruth Wu.

Published/Created	1990.
Description	vii, 166 leaves: ill.
LC classification	Microfilm 92/656 (Q)
Dissertation note	Thesis (Ph. D.)--University of Chicago, 1990.
Additional formats	Microfilm. Chicago, Ill.: University of Chicago, Joseph Regenstein Library, Dept. of Photoduplication, 1990. 1 microfilm reel; 35 mm.

Histochemical demonstration of tryptamine; properties of the formaldehyde-induced fluorophores of tryptamine and related indole compounds in models,

LCCN	72853110
Type of material	Book
Personal name	Björklund, Anders, 1945-
Main title	Histochemical demonstration of tryptamine; properties of the formaldehyde-induced fluorophores of tryptamine and related indole compounds in models, by Anders Björklund, Bengt Falck and Rolf Håkanson.
Published/Created	Lund, 1968.
Description	31 p. illus. 23 cm.
LC classification	QP801.A48 B53
Related names	Falck, Bengt, joint author. Håkanson, Rolf, joint author.
Subjects	Tryptamine. Fluorescence microscopy. Histochemistry.
Notes	Bibliography: p. 30-31.
Series	Acta physiologica Scandinavica. Supplementum 318

Histological responses of cabbage plants grown at different levels of nitrogen nutrition to indole (3) acetic acid ...

LCCN	a 42003999

Type of material	Book
Personal name	Goldberg, Ethel, 1910- [from old catalog]
Main title	Histological responses of cabbage plants grown at different levels of nitrogen nutrition to indole (3) acetic acid ... by Ethel Goldberg...
Published/Created	[n.p., 1938]
Description	1 p.l., 347-369 p. illus. 24 cm.
LC classification	QK753.I5 G6
Subjects	Growth (Plants) Indoleacetic acid. Plants--Effect of nitrogen on Cabbage.

I. Boron trifluoride as a condensing agent in the Fischer indole synthesis. II.

LCCN	a 44000752
Type of material	Book
Personal name	Smith, Curtis William, 1918- [from old catalog]
Main title	I. Boron trifluoride as a condensing agent in the Fischer indole synthesis. II. Carbon-alkylation with quaternary ammonium salts. A new synthesis of tryptophan and other compounds containing the 3-indolemethylene group.
Published/Created	Urbana, Ill., 1943.
Description	1 p.l., 3, [1] p. 23 cm.
LC classification	QD181.B1 S5
Subjects	Boron trifluoride. Indole. Ammonium salts. Tryptophan.

I. Studies of the reactions of substituted diketopiperazines. II.

LCCN	a 46000356
Type of material	Book

Personal name	Rowland, Ralph Lawrence, 1921- [from old catalog]
Main title	I. Studies of the reactions of substituted diketopiperazines. II. A study of the berlinerblau indole synthesis.
Published/Created	Urbana, Ill., 1945.
Description	1 p.l., 3, [1] p. 23 cm.
LC classification	QD305.H7 R64
Subjects	Piperazine. Indole.

Indolalkaloide.

LCCN	73084679
Type of material	Book
Personal name	Hesse, Manfred, 1935-
Main title	Indolalkaloide.
Published/Created	[Weinheim/Bergstr.] Verlag Chemie [c1974]
Description	2 v. illus. 25 cm.
ISBN	3527254919 (set)
LC classification	QC463.I5 H47
Contents	T. 1. Text.--T. 2. Spektren.
Subjects	Indole alkaloids--Spectra. Mass spectrometry--Charts, diagrams, etc.
Notes	Includes bibliographies.
Series	Fortschritte der Massenspektrometrie, v. 1 Progress in mass spectrometry, v. 1.

Indole alkaloids; an introduction to the enamine chemistry of natural products,

LCCN	65026890
Type of material	Book
Personal name	Taylor, William I.
Main title	Indole alkaloids; an introduction to the enamine chemistry of natural products, by W.I. Taylor.

Edition	[1st ed.].
Published/Created	Oxford, New York, Pergamon Press [1966]
Description	xi, 148 p. illus. 20 cm.
LC classification	QD421 .T28
Subjects	Indole alkaloids.
Notes	Includes bibliographies.
Series	The Commonwealth and international library of science, technology, engineering, and liberal studies. A Course in organic chemistry, advanced section, v. 27

Indole and biogenetically related alkaloids

LCCN	79042802
Type of material	Book
Main title	Indole and biogenetically related alkaloids / edited by J. D. Phillipson and M. H. Zenk.
Published/Created	London; New York: Academic Press, 1980.
Description	xvii, 379 p.: ill.; 24 cm.
ISBN	0125544502
LC classification	QK898.I53 I53 1980
Related names	Phillipson, J. D. (John David) Zenk, M. H., 1933- Phytochemical Society of Europe.
Subjects	Indole alkaloids--Congresses. Botanical chemistry--Congresses. Apocynaceae--Congresses. Loganiaceae--Congresses. Rubiaceae--Congresses. Plant chemotaxonomy--Congresses.
Notes	Based on lectures presented at an international symposium arranged by the Phytochemical Society of Europe in April 1979. Includes bibliographies and indexes.

Series	Phytochemical Society of Europe series; no. 17 0309-9393

Indole ring synthesis: from natural products to drug discovery

LCCN	2016007851
Type of material	Book
Personal name	Gribble, Gordon W.
Main title	Indole ring synthesis: from natural products to drug discovery / Gordon W. Gribble, Department of Chemistry, Dartmouth College, USA.
Published/Produced	Chichester, West Sussex: Wiley, 2016.
Description	xv, 688 pages: illustrations; 29 cm
ISBN	9780470512180 (cloth)
LC classification	QD400.5.S95 G75 2016
Subjects	Organic compounds--Synthesis. Indole. Ring formation (Chemistry) Transition metal catalysts.
Notes	Includes bibliographical references and index.
Additional formats	Online version: Gribble, Gordon W. Indole ring synthesis Chichester, West Sussex: John Wiley & Sons, Inc., 2016 9781118683057 (DLC) 2016010584

Indole.

LCCN	82201956
Type of material	Book
Personal name	Kudchadker, S. A.
Main title	Indole.
Published/Created	Washington, D.C. (2101 L St., N.W., Washington 20037): American Petroleum Institute, c1982.
Description	vi, 43 p.; 28 cm.
LC classification	QD401 .K97125 1982
Related names	Wilhoit, R. C.

Subjects	Indole--Tables.
Notes	"Written by S.A. Kudchadker and R.C. Wilhoit"--P. iii.
	"April 1982."
	Includes bibliographical references.
Series	API monograph series
	Publication / American Petroleum Institute; 719

Indole-3-aldehyde and certain of its condensation products

LCCN	a 46001427
Type of material	Book
Personal name	Van Order, Robert Bruce, 1915- [from old catalog]
Main title	Indole-3-aldehyde and certain of its condensation products
Published/Created	[n.p., 1945]
Description	128-133, 69-96 p. incl. tables. 25 cm.
LC classification	QD401 .V25
Related names	Lindwall, Harry Gustave, 1902- [from old catalog] joint author.
Subjects	Indole.
	Aldehydes.
	Condensation products (Chemistry)

Indoles.

LCCN	76154323
Type of material	Book
Main title	Indoles. Edited by William J. Houlihan. Contributors, William A. Remers [and others].
Published/Created	New York, Wiley-Interscience [1971-79, c1972-79]
Description	3 v. illus. 23 cm.
ISBN	0471375004 (v. 1)
LC classification	QD401 .I4

Related names	Houlihan, William J., 1930- ed. Remers, William A. (William Alan), 1932-
Subjects	Indole.
Notes	Includes bibliographical references.
Series	The Chemistry of heterocyclic compounds, v. 25

Instrumental data for drug analysis

LCCN	2005048542
Type of material	Book
Main title	Instrumental data for drug analysis / by Terry Mills, III ... [et al.].
Edition	3rd ed.
Published/Created	Boca Raton, FL: CRC Press, 2006.
Description	6 v. (xv, 4590 p.): ill.; 29 cm.
Links	Publisher description http://www.loc.gov/catdir/enhancements/fy0646/2005048542-d.html
ISBN	9780849319747 (alk. paper) 0849319749 (alk. paper)
LC classification	RS189 .M54 2006
Related names	Mills, Terry. Instrumental data for drug analysis.
Contents	V. 1. Acebutolol -- Dapsone -- v. 2. Dauromycin -- Indigotin Disulfonate -- v. 3. Indole-3-Acetic Acid -- Phendimetrazine -- v. 4. Phenelzine -- Zoxazolamine -- v. 5. Raman spectra data -- Mass spectra of pesticides -- Mass spectra of derivatized drugs -- Gas chromatography/infrared spectra of drugs -- v. 6. Standard KBr infrared spectra and standard NMR solvent spectra -- Supplemental infrared spectra -- Supplemental NMR spectra -- Ultraviolet absorption maxima -- Infrared index -- Mass spectra indices of drugs -- Mass spectra indices of pesticides -- Molecular formula index -- Gas chromatographic data -- Index to specific compounds.

Subjects	Drugs--Analysis--Standards. Drugs--Spectra--Standards. Instrumental analysis--Standards.
Notes	Includes index. Previous ed.: Instrumental data for drug analysis / Terry Mills. Boca Raton: CRC Press, 1993-1996.

Isoprenoid synthesis in plants and microorganisms: new concepts and experimental approaches

LCCN	2012946735
Type of material	Book
Main title	Isoprenoid synthesis in plants and microorganisms: new concepts and experimental approaches / Thomas J. Bach, Michel Rohmer, editors.
Published/Produced	New York, NY: Springer, [2013] ©2013
Description	xix, 505 pages: illustrations (some color); 26 cm
ISBN	9781461440628 (alk. paper) 1461440629 (alk. paper) 9781461440635 (ebk.)
LC classification	QK898.I76 I86 2013
Related names	Bach, Thomas J., editor of compilation. Rohmer, Michel, editor of compilation.
Contents	Isoprenoid Biosynthesis in Prokaryotic Organisms / Manuel Rodríguez-Concepción and Albert Boronat -- 1-Deoxy-d-Xylulose 5-Phosphate Synthase (DXS), a Crucial Enzyme for Isoprenoids Biosynthesis / Song Xiang, Gerlinde Usunow, Gudrun Lange, Marco Busch and Liang Tong -- Biosynthetic Genes and Enzymes of Isoprenoids Produced by Actinomycetes / Tohru Dairi -- Interactions of Isoprenoid Pathway Enzymes and Indirect Stimulation of Isoprenoid

Biosynthesis by Pentose Phosphate Cycle Substrates in Synechocystis PCC 6803 / Kelly Poliquin, Francis X. Cunningham Jr., R. Raymond Gantt and Elisabeth Gantt -- Metabolic Engineering of Monoterpenoid Production in Yeast / Marc Fischer, Sophie Meyer, Maryline Oswald, Patricia Claudel and Francis Karst -- Metabolic Engineering of Isoprenoid Production: Reconstruction of Multistep Heterologous Pathways in Tractable Hosts / Jérôme Maury, Mohammad A. Asadollahi, Luca R. Formenti, Michel Schalk and Jens Nielsen.

Microbially Derived Semisynthetic Artemisinin / Christopher J. Paddon, Derek McPhee, Patrick J. Westfall, Kirsten R. Benjamin and Douglas J. Pitera, et al. -- Artemisinin: Controlling Its Production in Artemisia annua / Pamela Weathers, Melissa Towler, Yi Wang and Kristin K. Wobbe -- Fosmidomycin as an Antimalarial Agent / Jochen Wiesner, Armin Reichenberg, Martin Hintz, Regina Ortmann and Martin Schlitzer, et al. -- Regulation of Isoprene and Monoterpene Emission / Isabel Nogués and Francesco Loreto -- Involvement of Compartmentalization in Monoterpene and Sesquiterpene Biosynthesis in Plants / Michael Gutensohn, Dinesh A. Nagegowda and Natalia Dudareva -- Strategies for the Manipulation of Carbocations by Aristolochene Synthase / David J. Miller and Rudolf K. Allemann -- Elucidating the Formation of Geranyllinalool, the Precursor of the Volatile C16-Homoterpene TMTT Involved in Indirect Plant Defense / Marco Herde, Katrin Gärtner, Tobias Köllner, Benjamin Fode and Wilhelm

Boland, et al. Strigolactones: A Cry for Help Results in Fatal Attraction. Is Escape Possible? / Carolien Ruyter-Spira, Juan Antonio López-Ráez, Catarina Cardoso, Tatsiana Charnikhova and Radoslava Matusova, et al. -- Prenyldiphosphate Synthases and Gibberellin Biosynthesis / Chris C.N. van Schie, Michel A. Haring and Robert C. Schuurink -- Gibberellin Phytohormone Metabolism / Reuben J. Peters -- Control of Plastidial Isoprenoid Precursor Supply: Divergent 1-Deoxy-d-Xylulose 5-Phosphate Synthase (DXS) Isogenes Regulate the Allocation to Primary or Secondary Metabolism / Michael H. Walter, Daniela S. Floss, Heike Paetzold, Kerstin Manke and Jessica Vollrath, et al. -- Tobacco Trichomes as a Platform for Terpenoid Biosynthesis Engineering / Alain Tissier, Christophe Sallaud and Denis Rontein -- Prenylated Proteins Are Required for Methyl-Jasmonate-Induced Monoterpenoid Indole Alkaloids Biosynthesis in Catharanthus roseus / Vincent Courdavault, Marc Clastre, Andrew John Simkin and Nathalie Giglioli-Guivarc'h. The Role of Prenylcysteine Methylation and Metabolism in Abscisic Acid Signaling in Arabidopsis thaliana / Dring N. Crowell and David H. Huizinga -- What We Do and Do Not Know About the Cellular Functions of Polyisoprenoids / Liliana Surmacz and Ewa Swiezewska -- Regulation of 3-Hydroxy-3-Methylglutaryl-CoA Synthase and 3-Hydroxy-3-Methylglutaryl-CoA Reductase and Rubber Biosynthesis of Hevea brasiliensis (B.H.K.) Mull. Arg / Pluang Suwanmanee, Nualpun Sirinupong and Wallie Suvachittanont -- Development of

Crops to Produce Industrially Useful Natural Rubber / Maureen Whalen, Colleen McMahan and David Shintani -- Occurrence of Two Acetoacetyl-Coenzyme A Thiolases with Distinct Expression Patterns and Subcellular Localization in Tobacco / Laurent Wentzinger, Esther Gerber, Thomas J. Bach and Marie-Andrée Hartmann. The Sterol C4-Demethylation in Higher Plants / Alain Rahier, Sylvain Darnet, Florence Bouvier and Bilal Camara -- Sterol C22-Desaturase and Its Biological Roles / Daisaku Ohta and Masaharu Mizutani -- Enzymatic Synthesis of Unnatural Cyclic Triterpenes / Ikuro Abe -- Saponin Synthesis and Function / Sam T. Mugford and Anne Osbourn -- Cardenolide Aglycone Formation in Digitalis / Wolfgang Kreis and Frieder Müller-Uri -- Biosynthesis of Isoprenoid Precursors in Arabidopsis / Manuel Rodríguez-Concepción, Narciso Campos, Albert Ferrer and Albert Boronat -- Understanding the Mechanisms that Modulate the MEP Pathway in Higher Plants / Patricia León and Elizabeth Cordoba -- Functional Analysis of HMG-CoA Reductase and Oxidosqualene Cyclases in Arabidopsis / Toshiya Muranaka and Kiyoshi Ohyama -- Systems Understanding of Isoprenoid Pathway Regulation in Arabidopsis / Eva Vranová.

Subjects Isopentenoids--Synthesis.

Notes Includes bibliographical references and index.

Jet spectroscopy of the indole chromophore in peptides using laser desorption

LCCN 92955254

Type of material Book

Personal name	Tubergen, Michael John, 1964-
Main title	Jet spectroscopy of the indole chromophore in peptides using laser desorption [microform] / by Michael John Tubergen.
Published/Created	1991.
Description	x, 254 leaves: ill.
LC classification	Microfilm 92/508 (Q)
Dissertation note	Thesis (Ph. D.)--University of Chicago, 1991.
Additional formats	Microfilm. Chicago, Ill.: University of Chicago, Joseph Regenstein Library, Dept. of Photod-uplication, 1992. 1 microfilm reel; 35 mm.

Kratom and other mitragynines: the chemistry and pharmacology of opioids from a non-opium source

LCCN	2014033470
Type of material	Book
Main title	Kratom and other mitragynines: the chemistry and pharmacology of opioids from a non-opium source / edited by Robert B. Raffa; assistant editors Jaclyn R. Beckett, Vivek N. Brahmbhatt, Theresa M. Ebinger, Chrisjon A. Fabian, Justin R. Nixon, Steven T. Orlando, Chintan A. Rana, Ali H. Tejani and Robert J. Tomazic.
Published/Produced	Boca Raton: CRC Press/Taylor & Francis Group, 2015.
Description	xv, 350 pages: illustrations; 24 cm.
ISBN	9781482225181 (hardcover: alk. paper)
LC classification	RM666.O6 K73 2015
Related names	Raffa, Robert B., editor.
Contents	Why do plants contain biologically active compounds? / Oné Pagán -- Short overview of mitragynines / Robert B. Raffa, Jaclyn R. Beckett, Vivek N. Brahmbhatt, Theresa M. Ebinger, Chrisjon A. Fabian, Justin R. Nixon, Steven T.

Orlando, Chintan A. Rana, Ali H. Tejani, and Robert J. Tomazic -- Psychopharmacological indole alkaloids from terrestrial plants / Allan Patrick G. Macabeo -- Salvinorin A: example of a non-alkaloidal bioactive opioid from a plant source / Jordan K. Zjawiony -- Mitragyna speciosa (Korth.) Havil / Sasha W. Eisenman -- Phytochemistry of Mitragyna speciosa / Vedanjali Gogineni, Francisco Leon, Bonnie A. Avery, Christopher McCurdy, and Stephen J. Cutler -- Chemistry of mitragynines: a brief introduction / Robert B. Raffa -- Chemistry of mitragyna compounds / Mariko Kitajima and Hiromitsu Takayama -- Medicinal chemistry based on mitragynine / Hiromitsu Takayama and Mariko Kitajima -- Chemical structures of opioids / Robert B. Raffa -- The detection of mitragynine and its analogs / Ruri Kikura-Hanajiri -- The SDME of mitragynine and analogs / Annette Cronin, Ines A. Ackerman, and Jochen Beyer -- Analgesic effects of mitragynine and analogs / Kenjiro Matsumoto and Syunji Horie -- Non-analgesic CNS effects / Jaclyn R. Beckett, Justin R. Nixon, and Ali H. Tejani -- Other opioid-associated endpoints / Steven T. Orlando and Vivek N. Brahmbhatt -- The kratom experience from firsthand reports / Earth Erowid and Fire Erowid -- The toxicology of indole alkaloids / Jennifer l. Ingram-Ross -- Opioid-induced adverse effects and toxicity / Joseph V. Pergolizzi, Jr. -- Toxicology of mitragynine and analogs / Surash Ramanathan and Sharif M. Mansor -- The use of animal models to measure tolerance, dependence, and abuse potential / Scott M. Rawls

-- Kratom: the epidemiology, use and abuse, addiction potential, and legal status / Zoriah Binti Aziz.

Subjects Analgesics, Opioid--chemistry.
Mitragyna--chemistry.
Analgesics, Opioid--pharmacology.
Opioid-Related Disorders--prevention & control.
Plant Preparations--chemistry.
Plant Preparations--pharmacology.

Notes Includes bibliographical references and index.

Laboratory technique. Ring test for indol

LCCN 10032321
Type of material Book
Personal name Grubbs, Samuel Bates, 1871- [from old catalog]
Main title Laboratory technique. Ring test for indol [by] S.B. Grubbs and Edward Francis. Collodium sacs [by] S.B. Grubbs and Edward Francis. Microphotography with simple apparatus [by] H.B. Parker. M.J. Rosenan ... director Hygienic laboratory.
Published/Created Washington, Govt. print. off., 1902.
Description 10 p. 1 l. illus., 2 pl. 23 cm.
LC classification RA421 .U4
LC copy QR66 .G8
Related names Francis, Edward, 1872- [from old catalog]
Parker, Herman B.
Subjects Indole.
Collodion.
Photomicrography.
Notes At head of title: ... Treasury department, U.S. Marine-hospital service Walter Wyman, surgeon-general.
Series United States. Hygienic laboratory. Bulletin no. 7

Lead optimization for medicinal chemists: pharmacokinetic properties of functional groups and organic compounds

LCCN	2012405828
Type of material	Book
Personal name	Dörwald, Florencio Zaragoza.
Main title	Lead optimization for medicinal chemists: pharmacokinetic properties of functional groups and organic compounds / Florencio Zaragoza Dorwald.
Published/Created	Weinheim: Wiley-VCH, c2012.
Description	xvi, 605 p.: ill.; 25 cm.
ISBN	9783527332267 (hbk.: alk. paper) 352733226X (hbk.: alk. paper) 9783527645664 (ePDF) 3527645667 (ePDF) 9783527645640 (oBook) 3527645640 (oBook) 9783527645657 (ePub) 3527645659 (ePub) 9783527645671 (mobi) 3527645675 (mobi)
LC classification	RS403 .D62 2012
Summary	"Small structural modifications can significantly affect the pharmacokinetic properties of drug candidates. This book, written by a medicinal chemist for medicinal chemists, is a comprehensive guide to the pharmacokinetic impact of functional groups, the pharmacokinetic optimization of drug leads, and an exhaustive collection of pharmacokinetic data, arranged according to the structure of the drug, not its target or indication. The historical origins of most drug classes and general aspects of modern drug

discovery and development are also discussed. The index contains all the drug names and synonyms to facilitate the location of any drug or functional group in the book. This compact working guide provides a wealth of information on the ways small structural modifications affect the pharmacokinetic properties of organic compounds, and offers plentiful, fact-based inspiration for the development of new drugs. This book is mainly aimed at medicinal chemists, but may also be of interest to graduate students in chemical or pharmaceutical sciences, preparing themselves for a job in the pharmaceutical industry, and to healthcare professionals in need of pharmacokinetic data." -- Publisher description.

Contents Part I: Introduction. The drug discovery process; lead optimization -- Part II: The pharmacokinetic properties of compound classes. Alkanes; Alkenes and alkynes; Arenes; Halides; Azides; Nitro compounds; Azo compounds; Triazenes; Nitrates and nitrites; N-nitroso compounds; N-oxides; Alcohols; Phenols; Ethers; Epoxides; Peroxides; Thiols; Thioethers; Sulfoxides; Sulfones; Aliphatic amines; Quaternary ammonium salts; Amidines; Guanidines, acylguanidines, and biguanides; Anilines; Hydrazines, acylhydrazines, and hydrazones; Aldehydes; Ketones; Carboxylic acids; Carboxylic esters; Amides; Lactams and imides; Nitriles; Carbonates; Carbamates; Ureas; Thiocarbonyl compounds; Sulfonic acids; Sulfonic esters; Sulfates And sulfamic acids; Phosphonic acids; Phosphoric acid derivatives; N-(aminoalkyl)benzamides, -

benzoates, and related compounds; Arylalkylamines; Phenethylamines (2-phenylethylamines); Aminoalkylindoles and indole alkaloids; Phenothiazines; Dibenzazepines and related tricyclic compounds; 3-aryloxy-2-hydroxypropylamines (b-adrenergic antagonists; 'b-blockers'); Opiates; N-(carboxyalkyl)-a-amino acid amides (prils); Anilides and amides of glycine; Peptides, peptidomimetics, and related oligoamides; Oligoarylamines, oligoarylamides, oligoarylcarbamates, and oligoarylureas; Imidazoles; Triazoles; Pyridines, pyrimidines, and related compounds; Quinolines; Nucleoside analogs; Dihydropyridines; Arenesulfonamides; Sulfonylureas; Benzodiazepines; Steroids; Anthracyclines; Arylacetic, benzoic, and related carboxylic acids (NSAIDS); Quinolonecarboxylic acids (gyrase inhibitors); B-lactams; Prostaglandin analogs; Sartans; Statins; Folic acid analogs (antifolates); Taxanes; Macrocyclic compounds.

Subjects Pharmaceutical chemistry.
Pharmacokinetics.
Drug Design.
Organic Chemicals--pharmacokinetics.
Pharmaceutical chemistry.

Notes Includes bibliographical references and index.

Lipoic acid: energy production, antioxidant activity and health effects

LCCN 2007035854

Type of material Book

Main title	Lipoic acid: energy production, antioxidant activity and health effects / edited by Mulchand S. Patel and Lester Packer.
Published/Created	Boca Raton: CRC Press, c2008.
Description	xxiii, 532 p.: ill.; 24 cm.
Links	Table of contents only http://www.loc.gov/catdir/toc/ecip0726/2007035854.html Publisher description http://www.loc.gov/catdir/enhancements/fy0742/2007035854-d.html
ISBN	9781420045376 (hardcover: alk. paper) 1420045377 (hardcover: alk. paper)
LC classification	QP772.T54 L49 2008
Related names	Patel, Mulchand S. Packer, Lester.
Contents	A trail of research on lipoic acid / Lester J. Reed -- Lipoic acid biosynthesis / Natasha M. Nesbitt ... [et al.] -- The search for potent alpha-lipoic acid derivatives: chemical and pharmacological aspects / Moriya Ben Yakir, Yehoshua Katz-hendler, and Shlomo Sasson -- Novel indole lipoic acid derivatives: synthesis and their antioxidant effects / A. Selen Gurkan and Erdem Buyukbingol -- Alpha-keto acid dehydrogenase complexes and glycine cleavage system: their regulation and involvement in pathways of carbohydrate, protein, and fat metabolism / Robert A. Harris ... [et al.] -- Pyruvate dehydrogenase complex regulation and lipoic acid / Lioubov G. Korotchkina and Mulchand S. Patel -- Role of lipoyl domains in the function and regulation of the mammalian pyruvate dehydrogenase complex / Thomas E. Roche ... [et al.] -- Inactivation and inhibition of alpha-ketoglutarate dehydrogenase: oxidative modification of lipoic acid / Kenneth M.

Humphries, Amy C. Nulton-Persson, and Luke I. Szweda -- Lipoate-protein ligase A: structure and function / Kazuko Fujiwara ... [et al.] -- An evaluation of the stability and pharmacokinetics of R-lipoic acid and R-dihydrolipoic acid dosage forms in human plasma from healthy subjects / David A. Carlson ... [et al.] -- Pharmacokinetics, metabolism, and renal excretion of alpha-lipoic acid and its main metabolites in humans / Jens Teichert and Rainer Preiss -- Modulation of cellular redox and metabolic status by lipoic acid / Derrick Han ... [et al.] -- Redoxin connection of lipoic acid / José Antonio Bárcena ... [et al.] -- Lipoic acid as an inducer of phase II detoxification enzymes through activation of Nrf2-dependent gene expression / Kate Peterson Shay, Swapna Shenvi, and Tory M. Hagen -- Deficiency disorders of components of PDH complex: E2, E3, and E3BP deficiencies / Jessie M. Cameron, Mary C. Maj, and Brian H. Robinson -- Relationship between primary biliary cirrhosis and lipoic acid / Carlo Selmi ... [et al.] -- Effects of lipoic acid on insulin action in animal models of insulin resistance / Erik J. Henriksen and Stephan Jacob -- Activation of cytoprotective signaling pathways by alpha-lipoic acid / Alexandra K. Kiemer and Britta Diesel -- Selenotrisulfide derivatives of alpha-lipoic acid: potential use as a novel topical antioxidant / William T. Self -- Alpha-lipoic acid: a potent mitochondrial nutrient for improving memory deficit, oxidative stress, and mitochondrial dysfunction / Jiankang Liu -- Effects of alpha-lipoic acid on AMP-activated protein kinase in

	different tissues: therapeutic implications for the metabolic syndrome / Eun Hee Koh ... [et al.].
Subjects	Lipoic acid.
	Thioctic acid--Health aspects.
	Thioctic acid--Therapeutic use.
	Thioctic acid--metabolism.
	Thioctic acid--pharmacology.
Notes	Includes bibliographical references and index.
Series	Oxidative stress and disease; 24

Monoterpenoid indole alkaloids. Supplement to Part 4

LCCN	94004941
Type of material	Book
Main title	Monoterpenoid indole alkaloids. Supplement to Part 4 / edited by J.E. Saxton.
Published/Created	Chichester [England]; New York: Wiley, c1994.
Description	xii, 848 p.: ill.; 24 cm.
Links	Publisher description http://www.loc.gov/catdir/description/wiley033/94004941.html
	Table of Contents http://www.loc.gov/catdir/toc/onix04/94004941.html
ISBN	0471951129
LC classification	QP801.I45 I53 1983 Suppl.
Related names	Saxton, J. Edwin (John Edwin), 1927-
Subjects	Indole alkaloids.
	Monoterpenes.
Notes	Includes bibliographical references and indexes.
Series	The Chemistry of heterocyclic compounds; v. 25, pt. 4

Notes on the presence of indol in sea foods and other food products

LCCN	26027064
Type of material	Book
Personal name	Clough, Ray William, 1885-

Main title	Notes on the presence of indol in sea foods and other food products, by Ray William Clough, Oscar Eugene Shostrom, Ernest Dunbar Clark.
Published/Created	Seattle, Wash., University of Washington Press, 1925.
Description	1 p. l., p. 103-108. 26 cm.
LC classification	TX612.F5 C5
LC copy	SH19 .W3 vol.1, no. 5 Copy 2.
Related names	Shostrom, Oscar Eugene, joint author. Clark, Ernest Dunbar, 1886- joint author.
Subjects	Indole. Skatol. Seafood. Food.
Series	University of Washington publications in fisheries. v. 1, no. 5. Sept., 1925

Of such a nature: Índole

LCCN	2017031737
Type of material	Book
Personal name	Kozer, José, author.
Main title	Of such a nature: Índole / José Kozer; translated by Peter Boyle.
Published/Produced	Tuscaloosa: University of Alabama Press, 2018.
Description	xxxi, 203 pages; 23 cm
ISBN	9780817359058 (pbk.) 9780817391676 (e-book)
LC classification	PQ7390.K68 I5213 2018
Parallel title	Índole
Related names	Kozer, José. Poems. Selections. Kozer, José. Poems. Selections. English.
Notes	Poems in English and Spanish, translated from Spanish.

Organic synthesis: state of the art 2005-2007

LCCN	2011377925
Type of material	Book
Personal name	Taber, D. F. (Douglass F.), 1948-
Main title	Organic synthesis: state of the art 2005-2007 / Douglass F. Taber.
Published/Created	Hoboken, N.J.: Wiley, c2008.
Description	xi, 228 p.: ill.; 25 cm.
Links	Inhaltsverzeichnis. http://bvbr.bib-bvb.de:8991/F?func=service&doc_library=BVB01&doc_number=016507653&line_number=0001&func_code=DB_RECORDS&service_type=MEDIA
ISBN	9780470288498 (hbk.) 0470288493 (hbk.)
LC classification	QD262 .T282 2008
Contents	1. Synthesis of (--)- Littoralisone -- 2. Enantiocontrolled Construction of Oxygenated and Animated Stereogenic Centers -- 3. Catalytic Enantioselective Construction of Alkylated Stereogenic Centers -- 4. Catalytic Enantioselective Aldol/Mannich Reactions Leading to Extended Arrays of Stereogenic Centers -- 5. Alternative Strategies for the Construction of Extended Arrays of Stereogenic Centers -- 6. Synthesis of (--)- Avrainvillamide and (+)-Stephacidin B -- 7. Best Synthetic Methods: Oxidation -- 8. Best Synthetic Methods: Reduction -- 9. Selective Reactions of Alkenes -- 10. Synthesis of the Potent FBBP12 Ligand Antascomicin B -- 11. Best Synthetic Methods: Carbon-Carbon Bond Formation -- 12. Enantioselective Construction of Quaternary Centers -- 13. Pd-Mediated Arylation of Aromatic and Hetero Aromatic Rings -- 14. Overman Route to

Gelsemine -- 15. Stereoselective Construction of Oxygen Heterocycles -- 16. Enantioselective Construction of Naturally-Occurring Cyclic Ethers -- 17. Stereoselective Construction of Nitrogen Heterocycles -- 18. Stork Synthesis of (--)- Reserpine -- 19. Stereocontrolled Construction of Azacyclic Natural Products -- 20. Preparation of Benzene Derivatives -- 21. Preparation of Heteroaromatics -- 22. Preparation and Reactions of Carboxylic Acids, Esters and Amides -- 23. Boger Route to (--)- Vindoline -- 24. Protection of C-O and C-N -- 25. New Catalysts and Strategies for Alkene Metathesis -- 26. Creative Applications of Alkene and Alkyne Metathesis in Total Synthesis (+)-8-epi-Xanthatin, (+)-Longicin, Latrunculin A, and Garsubellin A -- 27. Synthesis of Erythronolide A -- 28. Best Synthetic Methods: C-C Bond Formation -- 29. Catalytic Enantioselective Homologation of Aldehydes to Alcohols and Amines -- 30. Catalytic Enantioselective Construction of Alkylated Stereogenic Centers -- 31. Enantioselective Construction of Arrays of Stereogenic Centers -- 32. Adventures in Complex Indole Synthesis: (--)-Fischerindole I, (+)-Fischerindole G and (+)-Weltwitindolinone A -- 33. New Dienes and Dienophiles for Intermolecular and Intramolecular Diels-Alder Cycloaddtions -- 34. Organocatalytic Preparation of Enantiomerically-Pure Carbocycles -- 35. Carbocycle Construction by the Opening of Strained Rings: Synthesis of Tremulenolide A -- 36. Corey Route to the Dolabellanes: Isoedunol and [beta]-Araneosene -- 37. Transition-Metal

Catalyzed Enantioselective Ring Construction -- 38. Best Synthetic Methods: Functional Group Transformation -- 39. Selective Reactions of Alkenes -- 40. Synthesis and Absolute Stereochemical Assignment of (--)-Galbulimima Alkaloid 13 -- 41. Preparation of Benzene Derivatives -- 42. Preparation of Heteroaromatic Derivatives -- 43. Functional Group Transformation -- 44. Functional Group Protection -- 45. Leighton Synthesis of Dolabelide D -- 46. Stereocontrolled Construction of N-Heterocycles -- 47. Stereocontrolled Construction of O-Heterocycles -- 48. Stereocontrolled Synthesis of O-Heterocyclic Natural Products -- 49. Ready Synthesis of (--)-Nigellamine A[subscript 2] -- 50. Advances in the Diels-Alder Reaction: Synthesis of (+)-Lycoridine and of Dolabellatrienone -- 51. Enantioselective Carbocyclic Construction -- 52. Transition Metal Mediated Carbocyclic Construction -- 53. Synthesis of (--)-Colombiasin A and (--)-Elisapterosin B -- 54. Enantioselective Synthesis of C-N Ring Containing Natural Products -- 55. New Catalysts and Strategies for Alkene and Alkyne Metathesis -- 56. Heterocyclic Natural Products by Alkene Metathesis -- 57. Carbobyclic Natural Products by Alkene Metathesis -- 58. Crimmins Synthesis of (+)-SCH 351448 -- 59. Enantioselective Construction of Alcohols and Amines -- 60. Enantioselective Construction of Alkylated Stereogenic Centers -- 61. Enantioselective Construction of Arrays of Stereogenic Centers -- 62. Sorensen Synthesis of (--)-Guanacastepene E -- 63. Selective Reactions of Alkenes -- 64. Best

Synthetic Methods: C-C Bond Formation -- 65. Functional Group Protection and Deprotection -- 66. Nicolaou Synthesis of Platensimycin -- 67. Stereocontrolled C-O Ring Construction -- 68. Stereocontrolled Natural Product Synthesis: Cyclic Ethers and Macrolides -- 69. Stereocontrolled C-N Ring Construction -- 70. Stereocontrolled Alkaloid Total Synthesis -- 71. Fukuyama Synthesis of Morphine -- 72. Oxidation and Reduction in Organic Synthesis -- 73. Interconversion of Organic Functional Groups -- 74. Best Synthetic Methods: Carbon-Carbon Bond Formation -- 75. Overman Synthesis of (--)-Sarain A -- 76. Recent Developments in Alkene Metathesis -- 77. Pushing the Limits of Alkene Metathesis in Natural Product Synthesis -- 78. Preparation of Benzene Derivatives -- 79. Padwa Synthesis of Aspidophytidine -- 80. Synthesis of Heteroaromatics -- 81. Enantioselective Construction of Alcohols and Amines -- 82. Enantioselective Construction of Alleviated Stereogenic Centers -- 83. Enantioselective Construction of Arrays of Stereogenic Centers -- 84. Gin Synthesis of Nominine -- 85. Intramolecular Diels-Alder Reaction in Natural Product Synthesis -- 86. Catalytic Enantioselective Carbon-Carbon Ring Construction -- 87. New Directions in C-C Ring Construction: The Overman Synthesis of Guanacastepene N -- 88. Pettus Synthesis of (+)-Rishirilide B -- 89. Selective Reactions of Alkenes -- 90. Selective C-H Functionalization -- 91. New Methods for Carbon-Carbon Bond Formation -- 92. Nakada Synthesis of (+)-Digitoxigenin -- 93. Preparation

	of Benzene Derivatives -- 94. Preparation of Heteroaromatics: The Movassaghi Synthesis of (+)-Chimonanthine -- 95. Organic Functional Group Transformation -- 96. Organic Functional Group Protection -- 97. Trost Synthesis of (--)-Terpestacin -- 98. Stereoselective C-N Ring Construction -- 99. Stereoselective C-O Ring Construction -- 100. Synthesis of (--)-Blepharocalyxin D, (--)-Lasonolide A, and Attenol A -- 101. Dark Synthesis of Vigulariol -- 102. Enantioselective Organocatalytic Synthesis of Carbocycles: The Iwabuchi Synthesis of (+)-Juvabione -- 103. Transition-metal Mediated Synthesis of Carbocycles: The Snapper Synthesis of Pleocarpenone -- 104. Enantioselective Construction of Carbocycles: The Williams Synthesis of (+)-Fusicoauritone -- 105. C-H Functionalization: The White Reagent.
Subjects	Organic compounds--Synthesis. Organic compounds--Synthesis--Research. Organische Synthese.
Notes	Includes indexes.
Additional formats	Online version: Taber, D. F. (Douglass F.), 1948- Organic synthesis. Hoboken, N.J.: Wiley, c2008 (OCoLC)645894869

Organometallics as catalysts in the fine chemical industry

LCCN	2012945424
Type of material	Book
Main title	Organometallics as catalysts in the fine chemical industry / volume editors: Matthias Beller, Hans-Ulrich Blaser; with contributions by H.-U. Blaser ... [et al.].
Published/Created	Berlin; New York: Springer, c2012.

Description	x, 154 p.: ill.; 25 cm.
ISBN	9783642328329 (alk. paper) 3642328326 (alk. paper)
LC classification	QD411 .O736 2012
Related names	Beller, Matthias. Blaser, H. U.
Contents	Palladium-Catalysed Coupling Reactions / Johannes G. de Vries -- Applications of Rhodium-Catalyzed Hydroformylation in the Pharmaceutical, Agrochemical, and Fragrance Industries / Gregory T. Whiteker and Christopher J. Cobley -- Ruthenium-Catalyzed Selective Hydrogenation for Flavor and Fragrance Applications / Philippe Dupau -- Asymmetric Hydrogenation / Hans-Ulrich Blaser, Benoi^t Pugin and Felix Spindler -- Case Study: Sequential Pd-Catalyzed Cross-Coupling Reactions; Challenges on Scale-Up / Ioannis N. Houpis -- Pilot Plant Scale Synthesis of an Aryl-Indole: Scale Up of a Suzuki Coupling / Adriano F. Indolese -- Development of a Mild and Robust Method for Palladium Catalyzed Cyanation on Large Scale / Per Ryberg -- Application of Ring-Closing Metathesis Strategy to the Synthesis of Vaniprevir (MK-7009), a 20-Membered Macrocyclic HCV Protease Inhibitor / Cheng-yi Chen. Organometallic compounds.
Subjects	Organometallic compounds. Catalysis.
Notes	Includes bibliographical references and index.
Series	Topics in organometallic chemistry, 1436-6002; 42

Peroxidases and catalases: biochemistry, biophysics, biotechnology, and physiology

LCCN	2009034000
Type of material	Book
Personal name	Dunford, H. Brian.
Main title	Peroxidases and catalases: biochemistry, biophysics, biotechnology, and physiology / H. Brian Dunford.
Edition	2nd ed.
Published/Created	Hoboken, N.J.: John Wiley & Sons, c2010.
Description	xv, 459 p.: ill.; 25 cm.
ISBN	9780470224762 (cloth)
LC classification	QP603.P4 D86 2010
Related names	Dunford, H. Brian. Heme peroxidases.
Contents	Historical: pioneering work on horseradish and yeast cytochrome C peroxidases -- Heme peroxidase and catalase families -- Horseradish peroxidase I: the native enzyme, compounds I and II, their structures, and their cycle -- Horseradish peroxidase II: two-electron reactions, ferrous peroxidase, compound III, the five oxidation states, oxygen evolution, and inactivation -- Horseradish peroxidase III: oscillations and peroxidase-oxidase reactions with nadh, indole-3-acetic acid, and isobutyraldehyde light emission -- Yeast cytochrome C peroxidase: reactions with small substrates -- Yeast cytochrome C peroxidase: reaction with cytochrome C -- Spectroscopy I: optical, resonance raman, and x-ray absorption -- Spectroscopy II: nuclear magnetic resonance, electron spin, and Mossbauer -- Theoretical -- Class I: ascorbate peroxidase -- Catalase-peroxidases and Mycobacterium tuberculosis -- Class II: lignin, manganese,

	versatile, and Coprinus cinereus peroxidases -- Other class III peroxidases -- Catalases -- Myeloperoxidase: enzymology -- Biomedical aspects of myeloperoxidase: halogenation reactions in cardiovascular disease, infection, and cancer -- Prostaglandin H synthase -- Thyroid peroxidase -- Lacto- and salivary peroxidases -- Chloroperoxidase from C. fumago -- Selenium-containing enzymes: glutathione peroxidase and iodothyronine deiodinase -- Structure and function of vanadium peroxidases -- Other heme peroxidases and enzymes -- Application of peroxidases.
Subjects	Peroxidase. Catalase. Peroxidases--chemistry. Peroxidases--physiology.
Notes	Rev. ed. of: Heme peroxidases / H. Brian Dunford. c1999. Includes bibliographical references and indexes.

Preparation and reactions of aliphatic nitro compounds containing an indole nucleus.

LCCN	a 53003185
Type of material	Book
Personal name	Pellegrini, John Peter, 1926- [from old catalog]
Main title	Preparation and reactions of aliphatic nitro compounds containing an indole nucleus.
Published/Created	Urbana, 1949.
Description	2 p. 23 cm.
LC classification	QD305.N8 P4
Subjects	Nitro compounds. Indole.

Regulation of terpenoid indole alkaloid biosynthesis in Catharanthus roseus: the tryptophan decarboxylase gene

LCCN	93108749
Type of material	Book
Personal name	Goddijn, Oscar Johannes Maria, 1963-
Main title	Regulation of terpenoid indole alkaloid biosynthesis in Catharanthus roseus: the tryptophan decarboxylase gene / Oscar Johannes Maria Goddijn.
Published/Created	Alblasserdam: Offsetdrukkerij Haveka BV, 1992.
Description	99 p.: ill.; 24 cm.
LC classification	QP613.A75 G63 1992
Subjects	Aromatic amino acid decarboxylases. Indole alkaloids--Metabolism. Catharanthus roseus--Metabolism. Aromatic amino acid decarboxykase genes.
Notes	Summary in Dutch. "Stellingen" ([1] leaf) inserted. Errata slip inserted. Includes bibliographical references.
Dissertation note	Thesis (doctoral)--Rijksuniversiteit te Leiden, 1992.

Stereoselective syntheses in the corynantheine series and of related model compounds

LCCN	96224107
Type of material	Book
Personal name	Tirkkonen, Birgit.
Main title	Stereoselective syntheses in the corynantheine series and of related model compounds / Birgit Tirkkonen.
Published/Created	Helsinki: Suomalainen Tiedeakatemia, 1994.
Description	33 p.: ill.; 25 cm.
ISBN	9514107578

LC classification QD1 .S785 no. 258
Subjects Stereochemistry.
Indole alkaloids--Synthesis.
Notes "Received 16 May 1994"--T.p. verso.
Includes bibliographical references (p. 31-33).
Series Annales Academiae Scientiarum Fennicae. Series A. II, Chemica, 0066-1961; 258

Studies of the metabolism of tryptophan labeled with N>15 in the indole ring.
LCCN a 51004000
Type of material Book
Personal name Schayer, Richard William, 1915- [from old catalog]
Main title Studies of the metabolism of tryptophan labeled with N>15 in the indole ring.
Published/Created New York, 1949 [i.e., 1950]
Description 10 p. tables. 23 cm.
LC classification QP551 .S38
Subjects Tryptophan.
Indole.

Studies of the synthesis of a thiophene analog of indole.
LCCN a 53003349
Type of material Book
Personal name Swaminathan, Sambasiva, 1923- [from old catalog]
Main title Studies of the synthesis of a thiophene analog of indole.
Published/Created Urbana, 1949.
Description 4 p. 23 cm.
LC classification QD401 .S98
Subjects Indole.
Aldehydes.

Studies on the mechanism of alkylation by Mannich bases of the indole series.

LCCN	a 53003398
Type of material	Book
Personal name	Eliel, Ernest Ludwig, 1921-
Main title	Studies on the mechanism of alkylation by Mannich bases of the indole series.
Published/Created	Urbana, 1948.
Description	4 p. 23 cm.
LC classification	QD281.A5 E4
Subjects	Alkylation. Indole.

Studies on the synthesis and mass spectra of aromatic organoselenium compounds, particularly indole derivatives.

LCCN	72432415
Type of material	Book
Personal name	Agenäs, Lars-Börge.
Main title	Studies on the synthesis and mass spectra of aromatic organoselenium compounds, partic-ularly indole derivatives.
Published/Created	Uppsala, Universitetet; Stockholm, Almqvist & Wiksell, 1969.
Description	29 p. 24 cm.
LC classification	Q64 .A63 no. 132
Subjects	Organoselenium compounds.
Notes	Bibliography: p. 27-29.
Dissertation note	Inaug. Diss.--Uppsala.
Series	Acta Universitatis Upsaliensis. Abstracts of Uppsala dissertations in science; 132.

Synthesis of indole compounds

LCCN	a 51006538
Type of material	Book
Personal name	Klioze, Oscar.
Main title	Synthesis of indole compounds / by Oscar Klioze.
Published/Created	1949.
Description	ii, 77 leaves: ill.; 29 cm.
LC classification	QD401 .K713
Subjects	Indole--Synthesis.
Notes	Typescript.
	Vita.
	Bibliography: leaves [72]-77.
Dissertation note	Thesis (Ph. D.)--University of Maryland, 1949.

Synthetic organic photochemistry

LCCN	84010480
Type of material	Book
Main title	Synthetic organic photochemistry / edited by William M. Horspool.
Published/Created	New York: Plenum Press, c1984.
Description	xvii, 534 p.: ill.; 23 cm.
ISBN	030641449X
LC classification	QD715 .S96 1984
Related names	Horspool, William M.
Contents	Photo addition and photocyclization processes of aromatic compounds / Andrew Gilbert -- Enone photochemical cycloaddition in organic synthesis / Alan C. Weedon -- Synthetic aspects of photochemical electron transfer reactions / Patrick S. Mariano and Jerome L. Stavinoha – Phthalimide and its derivatives / John D. Coyle -- Photochemical addition reactions in the benzo-(b)thiophene, benzo(b)furan and indole series / Douglas C. Neckers and Alois H.A. Tinnemans -

	- Azirine photolysis and cycloaddition reactions / Albert Padwa -- Photoremovable protecting groups / Roger W. Binkley and Thomas W. Flechtner -- Photochemical synthesis of oxetans / Howard A.J. Carless -- Equipment and techniques / William M. Horspool.
Subjects	Organic photochemistry.
	Organic compounds--Synthesis.
Notes	Includes bibliographical references and index.

The Aldrich library of NMR spectra,

LCCN	74077906
Type of material	Book
Personal name	Pouchert, Charles J.
Main title	The Aldrich library of NMR spectra, by Charles J. Pouchert and John R. Campbell.
Published/Created	[Milwaukee, Wis., Aldrich Chemical Co., 1974]
Description	11 v. illus. 31 cm.
LC classification	QD96.N8 P68
Related names	Campbell, John R., 1943- joint author.
Contents	v. 1. Non-aromatic hydrocarbons, halogenated hydrocarbons, alcohols, ethers, mercaptans, & sulfides.--v. 2. Non-aromatic amines, nitro & nitroso compounds, ketones, aldehydes & carboxylic acids.--v. 3. Non-aromatic amino acids, esters, lactones, anhydrides, amides, nitriles & cumulated double bonds.--v. 4. Aromatic hydrocarbons, halides, ethers, and phenols.--v. 5. Aromatic alcohols, mercaptans, sulfides, amines, nitro & nitroso compounds.--v. 6. Aromatic ketones, aldehydes, & carboxylic acids.--v. 7. Aromatic amino acids, esters, lactones, anhydrides, acid halides, amides, nitriles & cumulated double bonds.--v. 8. Aromatic 5-membered single

	and fused-ring heterocycles.--v. 9. Aromatic 6-membered single and fused ring heterocycles & N-oxides.--v. 10. Sulfur-oxygen compounds, phosphorous compounds, oximes, alkynes, steroids, indole alkaloids, silanes, boranes & miscellaneous compounds.--v. 11. Cumulative index.
Subjects	Nuclear magnetic resonance spectroscopy.

The catalytic dehydrogenation of piperidine.

LCCN	a 52000181
Type of material	Book
Personal name	Meyers, Cal Yale, 1927- [from old catalog]
Main title	The catalytic dehydrogenation of piperidine.
Published/Created	Urbana, 1951.
Description	3 p. 23 cm.
LC classification	QD401 .M627
Subjects	Piperidine. Indole.

The chemistry of indoles

LCCN	71107554
Type of material	Book
Personal name	Sundberg, Richard J., 1938-
Main title	The chemistry of indoles [by] Richard J. Sundberg.
Published/Created	New York, Academic Press, 1970.
Description	xi, 489 p. illus. 24 cm.
ISBN	0126769508
LC classification	QD401 .S95
Subjects	Indole.
Notes	Includes bibliographies.
Series	Organic chemistry; a series of monographs, v. 18 Organic chemistry (New York, N.Y.); v. 18.

The condensation of indole derivatives with aldehydes

LCCN	24019138
Type of material	Book
Personal name	Burr, George O. (George Oswald), 1896-
Main title	The condensation of indole derivatives with aldehydes
Published/Created	Easton, Pa., Eschenbach printing co., 1924.
Description	25, [1] p. diagrs. 24 cm.
LC classification	QD401 .B96
Subjects	Indole. Aldehydes. Humin. Proteins.

The effect of indole acetic acid and other growth promoting substances on the endogenous respiration of the Avena coleoptile.

LCCN	54033281
Type of material	Book
Personal name	Anker, Leendert.
Main title	The effect of indole acetic acid and other growth promoting substances on the endogenous respiration of the Avena coleoptile.
Published/Created	Amsterdam, North-Holland Pub. Co., 1953.
Description	65 p. diagrs. 25 cm.
LC classification	QK891 .A5
Subjects	Plants--Respiration. Indoleacetic acid. Growth factors.
Notes	Proefschrift--Utrecht. "Dit proefschrift verschijnt tevens in Acta botanica Neerlandica, vol. II, nr. I." "Stellingen": [2] p. inserted. Bibliography: p. 63-65.

The Fischer indole synthesis

LCCN	81014749
Type of material	Book
Personal name	Robinson, Brian, 1936-
Main title	The Fischer indole synthesis / Brian Robinson.
Published/Created	Chichester; New York: Wiley, c1982.
Description	xiii, 923 p.: ill.; 24 cm.
ISBN	0471100099:
LC classification	QD401 .R598 1982
Subjects	Indole. Organic compounds--Synthesis.
Notes	"A Wiley Interscience publication." Includes bibliographical references and index.

The Indoles: the monoterpenoid indole alkaloids

LCCN	82021958
Type of material	Book
Main title	The Indoles: the monoterpenoid indole alkaloids / edited by J. Edwin Saxton.
Published/Created	New York: Wiley, 1983.
Description	xii, 886 p.: ill.; 25 cm.
ISBN	0471897485:
LC classification	QP801.I45 I53 1983
Subjects	Saxton, J. Edwin (John Edwin), 1927- Indole alkaloids.
Notes	"An Interscience publication." Includes bibliographical references and indexes.
Series	The Chemistry of heterocyclic compounds; v. 25, pt. 4

The isocyano group as a carbenoid function cyclization of ortho-isocyanobiphenyl and derivatives.

LCCN	73153938
Type of material	Book

Personal name	Jong, Johannes de.
Main title	The isocyano group as a carbenoid function cyclization of ortho-isocyanobiphenyl and derivatives.
Published/Created	Groningen, V.R.B. Offsetdrukkerij, 1971.
Description	86 p. illus. 24 cm.
LC classification	QD603.C1 J66
Subjects	Isocyanides. Indole. Ring formation (Chemistry) Photochemistry.
Notes	Extra t.p. with thesis statement inserted. Summary in Dutch. "Stellingen": [2] p. inserted. Includes bibliographical references.
Dissertation note	Proefschrift--Rijksuniversiteit te Groningen.

The mechanism of the action of indole-3-acetic acid on the water absorption by Avena coleoptile sections.

LCCN	56035011
Type of material	Book
Personal name	Ketellapper, Hendrik Jan.
Main title	The mechanism of the action of indole-3-acetic acid on the water absorption by Avena coleoptile sections.
Published/Created	Amsterdam, North-Holland Pub. Co., 1953.
Description	388-444 p. illus., diagrs., tables. 25 cm.
LC classification	QK871 .K4
Subjects	Plants--Absorption of water. Indoleacetic acid.
Notes	Proefschrift--Utrecht. "Verschijnt tevens in Acta botanica neerlandica, vol. 2 (4), 1953." "Stellingen" ([2] p.) inserted.

Bibliography: p. 441-444.

The metabolism of indole and tryptophane derivatives,

LCCN	37004693
Type of material	Book
Personal name	Bauguess, Lyle Clyde, 1910- [from old catalog]
Main title	The metabolism of indole and tryptophane derivatives,
Published/Created	[Baltimore, 1936]
Description	1 p.l., p. 675-699, 615-623, 253-257. diagrs. 23 cm.
LC classification	QP551 .B28 1934
Related names	Berg, Clarence Peter, 1900- [from old catalog] joint author.
Subjects	Tryptophan. Indole. Metabolism.

The nature of the reaction of T2 bacteriophage with Indole.

LCCN	tmp93001821
Type of material	Book
Personal name	Kanner, Lee Chevlen, 1925- [from old catalog]
Main title	The nature of the reaction of T2 bacteriophage with Indole.
Published/Created	[n.p., n.d.]
Description	vi, 56 l. ill.
LC classification	Microfilm 8031
Notes	Bibliography: leaves 40-41.
Dissertation note	Thesis (Ph. D.)--University of Chicago, 1963.
Additional formats	Microfilm. Chicago [Library, Dept. of Photographic Reproduction, University of Chicago] 1963.

The preparation of some polymethylphenylhydrazones by the Japp-Klingemann reaction and the Fischer indole reaction of ethyl pyruvate 2,6-dimethylphenylhydrazone.

LCCN	55037285
Type of material	Book
Personal name	Henley, William O. [from old catalog]
Main title	The preparation of some polymethylphenylhydrazones by the Japp-Klingemann reaction and the Fischer indole reaction of ethyl pyruvate 2,6-dimethylphenylhydrazone.
Published/Created	Pittsburgh, 1954.
Description	viii, 87 l. diagrs., tables. 29 cm.
LC classification	QD305.A9 H57
Subjects	Hydrazones.

The synthesis and reactions of methoxyindole compounds.

LCCN	a 55007234
Type of material	Book
Personal name	Bell, John Barr, 1919- [from old catalog]
Main title	The synthesis and reactions of methoxyindole compounds.
Published/Created	[n.p., 1948]
Description	547-553 p. table. 26 cm.
LC classification	QD401 .B43
Related names	Lindwall, Harry Gustave, 1902- [from old catalog] joint author.
Subjects	Indole. Chemical reactions.

The synthesis of compounds related to indole alkaloids.

LCCN	a 52001360
Type of material	Book
Personal name	Lorenz, John Clark, 1923- [from old catalog]

Main title The synthesis of compounds related to indole alkaloids.
Published/Created Urbana, 1951.
Description 4 p. 23 cm.
LC classification QD421 .L88
Subjects Indole.
Yohimbine.
Anhydrides.

The triptans: novel drugs for migraine
LCCN 2001036505
Type of material Book
Main title The triptans: novel drugs for migraine / edited by Patrick Humphrey, Michel Ferrari, and Jes Olesen.
Published/Created Oxford; New York: Oxford University Press, 2001.
Description xv, 341 p.: ill.; 25 cm.
Links Table of contents only http://www.loc.gov/catdir/enhancements/fy0612/2001036505-t.html
Publisher description http://www.loc.gov/catdir/enhancements/fy0612/2001036505-d.html
ISBN 0192632140 (hbk: alk. paper)
LC classification RC392 .T75 2001
Related names Humphrey, Patrick.
Olesen, Jes.
Subjects Migraine--Chemotherapy.
Serotonin--Agonists--Therapeutic use.
Sumatriptan.
Indole alkaloids--Therapeutic use.
Migraine--drug therapy.
Serotonin Agonists--therapeutic use.
Notes Includes bibliographical references and index.
Series Frontiers in headache research; vol. 10

Trends in cancer prevention research

LCCN	2005036620
Type of material	Book
Main title	Trends in cancer prevention research / Frank Columbus (editor).
Published/Created	New York: Nova Science Publishers, c2007.
Description	xi, 116 p.; 27 cm.
Links	Table of contents only http://www.loc.gov/catdir/toc/ecip065/2005036620.html
ISBN	159454946X 9781594549465
LC classification	RC268 .T74 2007
Related names	Columbus, Frank H.
Contents	Increased incidence of cancer near a cell-phone transmitter station / Ronni Wolf and Danny Wolf -- Vitamin D and cancer chemoprevention / Rajendra G. Mehta -- A meta-analysis of soyfoods and risk of breast cancer in women 281 / Lin Yan and Edward Spitznagel -- The prevention of AIDS-related malignancies / Justin Stebbing ... [et al.] -- The role of specific estrogen metabolites in the initiation of breast and other human cancers / Eleanor G. Rogan and Ercole L. Cavalieri -- A growth hormone polymorphism at position 1663 in intron 4 of the GH1 gene is not associated with breast cancer risk / Eva S. Schernhammer ... [et al.] -- Recruiting African American women into chemoprevention trials: Gail model as an education tool / Karen Patricia Williams, Resche D. Hines, and Jonathan N. Livingston -- Effects of indole phytoalexins from cruciferous plants on the growth of cancer cells. Implications for cancer chemoprevention and chemotherapy / Roman Mezencev ... [et al.] -- Modulatory influence of

	biochanin A on cyclooxygenase -2 and inducible nitricoxide synthase expression in azoxymethane induced colonic aberrant crypts / Archana Sengupta, Shamee Bhattacharjee and Samit Ghosh.
Subjects	Cancer--Prevention.
	Neoplasms.
	Neoplasms--prevention & control.
Notes	Includes bibliographical references and index.

RELATED NOVA PUBLICATIONS

INDOLEAMINES: SOURCES, ROLE IN BIOLOGICAL PROCESSES AND HEALTH EFFECTS

Editor: Angel Catalá
(Instituto de Investigaciones Fisicoquímicas Teóricas y Aplicadas, (INIFTA-CCT La Plata-CONICET), Facultad de Ciencias Exactas, Universidad Nacional de La Plata, Argentina)

ISBN: 978-1-63482-097-4 Binding: Hardcover
Publication Date: 2015

This book presents an overview of the role in biological processes and health effects produced by indoleamines, which are ubiquitous and versatile molecules with important physiological functions. The topics analyzed in this book cover a broad spectrum of functions played by indoleamines including the role of indoleamines in reducing free radical damage and oxidative stress; melatonin; endogenous sources and role in the regulation of fish reproduction; the neuroimmune roles of indoleamine pathways in microorganisms also produce IAA. IAA might be used as a signaling molecule in microbial communication to induce the expression of genes involved in plant niche adaptation and colonization. It also plays a role in promoting plant cell proliferation and elongation. Interactions between IAA-producing microorganisms and plants lead to diverse outcomes on the plant side, varying from pathogenesis to phytostimulation. This review describes various routes adopted by microorganisms for IAA production and discusses past and recent data on IAA as signaling molecule.

Acetic Acids: Indole-3-Acetic Acid (IAA), Indole-3-Butyric Acid (IBA), A-Naphthaleneacetic Acid (A-NAA) and 2,4 Dichlorophenoxy Acetic Acid (2,4-D): Chemical Properties, Production and Applications in Tissue Culture, on Rooting and on Biochemical Parameters*

Virginia Sarropoulou,† Kortessa Dimassi-Theriou and Ioannis Therios

Laboratory of Pomology, Department of Horticulture, School of Agriculture, Aristotle University of Thessaloniki, Thessaloniki, Greece

In this chapter, the effects of the auxins IAA, IBA, a-NAA and 2,4-D from the category of the acetic acids in tissue culture, on rooting, on photosynthetic pigments (total leaf chlorophyll (a+b) and carotenoids content) as well as on total carbohydrate and proline concentration were analysed. Acetic acid is a weak organic compound with the chemical

---

* The full version of this chapter can be found in *Acetic Acids: Chemical Properties, Production and Applications*, edited by Angelo Basile, published by Nova Science Publishers, Inc, New York, 2013.

† Corresponding Author E-mail: vsarrop@gmail.com; vsarropo@agro.auth.gr.

formula CH_3CO_2H (also written as CH_3COOH or $C_2H_4O_2$) and is the main component of vinegar. Acetic acid is mildly corrosive to metals including iron, magnesium, and zinc, forming hydrogen gas and salts called acetates. At physiological pHs, acetic acid is usually fully ionized to acetate. In biochemistry, acetate and acetic acid are equivalent. Acetic acid is produced industrially both synthetically and by bacterial fermentation. Glacial acetic acid is an excellent polar protic solvent, is often used as a solvent for reactions involving carbocations, such as Friedel-Crafts alkylation and is used in analytical chemistry for the estimation of weakly alkaline substances such as organic amides. Auxins derive their name from the Greek word *auxein* - "to grow/increase".

The most important member of the auxin family is indole-3-acetic acid (IAA). Synthetic auxin analogs include 1-naphthaleneacetic acid, 2,4-dichlorophenoxyacetic acid (2,4-D), and many others. Auxins are also often used to promote initiation of adventitious roots, to promote uniform flowering and fruit set, and to prevent premature fruit drop. On the cellular level, auxin is essential for cell growth, affecting both cell division and cellular expansion. Auxin participates in phototropism, geotropism, hydrotropism and other developmental changes. Auxins promote root initiation, induce shoot apical dominance and delay fruit senescence. In low concentrations, auxins can inhibit ethylene formation and transport in plants; however, high concentrations can induce the synthesis of ethylene. IAA is chemically similar to the amino acid tryptophan which is generally accepted to be the molecule from which IAA is derived. 2,4-D is absorbed through the leaves and is translocated to the meristems of the plant. Auxins and their synthetic analogues are commonly used for rooting of cuttings, but their efficiency depends on experimental set-up and, even more importantly, on species or cultivar, and type of explants investigated. A favourable rooting effect of IBA may be due to its higher stability in comparison to IAA, its slow transport from the site of application at the base of cutting and its conversion to IAA in the cutting.

INDEX

G

H

I

Q

R

S

T

U

V

W

Y

Z

Chemical Reactions in Gas, Liquid and Solid Phases: Synthesis, Properties and Application

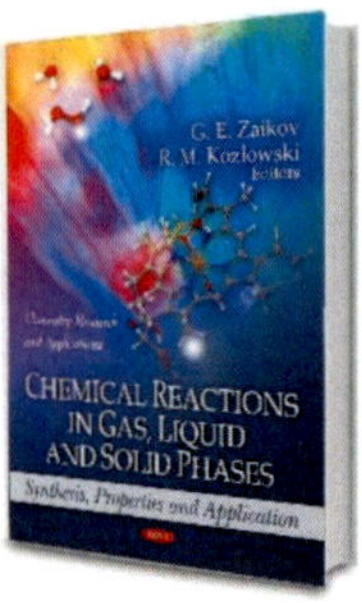

Editor: G.E. Zaikov (Institute of Biomedical Physics, Moscow, Russia), R.M. Kozlowski (ESCORENA Program, Poznan, Poland)

Series: Chemistry Research and Applications

Book Description: This new book examines topical data on the subject of chemical reactions in different phases of gas, liquid and solid states, such as the classification of polymers in reactivity toward nitrogen oxide; and the influence of the initiation rate of radicals on the kinetic characteristics of quercetin and dihydroquercetin in the methyl oleate oxidation.

Hardcover ISBN: 978-1-61668-671-0
Retail Price: $265

Indoleamines: Sources, Role in Biological Processes and Health Effects

Editors: Angel Catalá (Instituto de Investigaciones Fisicoquímicas Teóricas y Aplicadas, (INIFTA-CCT La Plata-CONICET), Facultad de Ciencias Exactas, Universidad Nacional de La Plata, Argentina)

Series: Biochemistry Research Trends

Book Description: This book presents an overview of the role in biological processes and health effects produced by indoleamines, which are ubiquitous and versatile molecules with important physiological functions. The topics analyzed in this book cover a broad spectrum of functions played by indoleamines including the role of indoleamines in reducing free radical damage and oxidative stress.

Hardcover ISBN: 978-1-63482-097-4
Retail Price: $215